The Invention of Mikhail Lomonosov

A Russian National Myth

IMPERIAL ENCOUNTERS IN RUSSIAN HISTORY

Series Editor: GARY MARKER
(State University of New York, Stony Brook)

ACADEMIC
STUDIES
PRESS

THE INVENTION
OF MIKHAIL LOMONOSOV
A Russian National Myth

S T E V E N A . U S I T A L O

B O S T O N / 2 0 1 3

Library of Congress Cataloging-in-Publication Data:
A catalog record for this book as available from the Library of Congress.

ISBN - 978-1-61811-806-6
ISBN - 978-1-61811-195-1 (electronic)

Book design by Ivan Grave

Published by Academic Studies Press in 2013
28 Montfern Avenue
Brighton, MA 02135, USA
press@academicstudiespress.com
www.academicstudiespress.com

To Margarita

Contents

ACKNOWLEDGEMENTS

This work has been in progress for a number of years; it is a particular pleasure, therefore, to finally express my appreciation to the many friends, colleagues, and organizations that have assisted me along the way. Paul Austin, Kees Boterbloem, Robert Collis, James Cracraft, James Delbourgo, Simon Dixon, the late Il'ia Serman, Marina Swoboda, the late Viktor Zhivov, and Ernest Zitser read portions of this work, primarily at its earliest stages. Their assessments were appropriately critical and encouraging. My work was made much easier due to the accommodating library staffs at the McClennan Library at McGill University and in St. Petersburg, at the Russian Academy of Sciences Library and at the National Library of Russia (Bronsilava Gradova was especially helpful at the latter institution). McGill University, Northern State University, the American Council for Teachers of Russian, and the Fulbright Program provided much-needed support to me over the years. Thanks to their generosity I was able to spend much time—perhaps too much time— in St. Petersburg.

My good friend and former office mate at McGill University, Ismail Rashid, first read this work when it began life as a dissertation. He has continued to read and critique it over the years. His always friendly advice and support remain one of my fondest memories of my time at McGill. Jeffrey Veidlinger listened patiently to many conference presentations based on this work, and has always been forthright and supportive. Colum Leckey, Kirill Ospovat, and Joachim Klein read the latest iterations of this volume, and I am indebted to each of them for their perceptive judgments. Ben Whisenhunt, my old friend from St. Petersburg and Chicago, read and re-read many versions of my manuscript and offered what is always most useful: encouragement. No one should have been forced to spend so much time reading my various efforts to make Lomonosov's life cohere. Ben, thank you!

Gary Marker buoyed me in a most effective manner, commisioning this work for the series he edits at Academic Studies Press. Gary's insightful appraisal of the work has decidedly improved the final product. Sharona Vedol was an ideal editor; in addition to shepherding the manuscript through publication, she responded to all my queries and concerns with alacrity. To Igor Nemirovsky, thank you for ensuring the acquisition and publication of *The Invention of Mikhail Lomonosov.*

I had the good fortune as a graduate student to study, principally, with Valentin Boss. Valentin was more than an encouraging advisor, later a friend; he inspired in me an enduring fascination with eighteenth-century Russian culture. Our discussions at the Lomonosov Museum were memorable. I could not have asked for a more interesting, erudite teacher. My late father Arnold and my mother Seija always provided me with what parents should: unconditional love and security. Words fail me—or almost fail me—when I think of my wife Margarita and daughter Izabella. Suffice it to say that I am profoundly happy, at times astonished, to have both of them in my life. Well, Margarita, now that Lomonosov is out of our life, it's off to our next adventure.

Versions of the introduction and some of the text were previously published. My thanks to the publishers for giving permission to partially reprint the following: "Russia's 'First' Scientist: The Self-Fashioning of Mikhail Lomonosov," in Steven A. Usitalo and William Benton Whisenhunt, eds., *Russian and Soviet History: From the Time of Troubles to the Collapse of the Soviet Union* (Rowman & Littlefield, 2008); and "Lomonosov: Patronage and Reputation at the St. Petersburg Academy of Sciences," *Jahrbücher für Geschichte Osteuropas*, 59, no. 2 (2011): 217-39.

INTRODUCTION

*F*or more than two hundred years the eighteenth-century polymath Mikhail Vasil'evich Lomonosov (1711-65) has been glorified in Russian culture as the "father" of Russian science, literature, and, more generally, learning.[1] The outlines of his biography are exceedingly familiar in his own country. Heroic tales describing the emergence of this son of a fisherman from the far northern periphery of Russia (he was born in a village not too distant from Arkhangel'sk, near the White Sea) were recited, albeit hardly voluntarily, by generations of Russian and Soviet schoolchildren. Lomonosov's indefatigable acquisition of knowledge, culminating in many productive years of activity at the St. Petersburg Academy of Sciences—conceived by Peter the Great, it remains to this day the fundamental scientific and cultural institution in Russia—became

---

[1]   While the origins of the idea of Lomonosov as the father, or founder, of Russian science and a "modernized" literature lies in the eighteenth century, with the birth of the myth of Lomonosov, as with so much else pertaining to the study of eighteenth- and nineteenth-century Russian cultural developments, the nineteenth-century social and literary critic Vissarion Belinskii seems to have given a more explicit, if now seemingly cliched, voice to already existing beliefs. Belinskii made extensive references to Lomonosov in his writings, and his pronouncements, always issued with an authoritative tone, were usually posed as aphorisms. To Belinskii, "Lomonosov was not only a poet, orator and litterateur, but a great scientist," someone who profoundly altered the lives of his compatriots by introducing the sciences and learning to Russia (the citation is from an 1836 review of Ksenofont Polevoi's two-volume historical novel on Lomonosov, printed in V. G. Belinskii, *Polnoe sobranie sochinenii*, vol. 2 [Moscow, 1953], 189). In a similar vein, he wrote that Lomonosov, who was quite unreservedly "brilliant" in his abilities, "is the father of Russian letters and learning" (from a short critique penned by Belinskii in 1844, in ibid., vol. 8 [Moscow, 1955], 359).

the stuff of legend. An accomplished physicist, chemist (his chair at the Academy of Sciences was in chemistry), poet, historian, linguist, geographer, artist, and more, he is the most celebrated personage identified with the Russian Enlightenment.[2]

Discussions over not only the nature, but also the very concept, of a Russian Enlightenment (*russkoe prosveshchenie*) at times became deeply colored by contemporary ideological dictates during the Soviet period. Especially with with the rise of a more assertive Russian nationalism in the 1940s and 1950s, many scholars began to insist that eighteenth-century Russian society experienced a rather expansive indigenous Enlightenment that at its apogee was marked by a thoroughgoing materialism. The more extreme political and social attributes that characterized said Russian Enlightenment were, however, never either universally accepted or even clearly delineated. Indeed, many studies that offered deeply researched monographic examinations of eighteenth-century Russian literary and cultural "links" (*sviazi*) with West European enlightenment thinkers also appeared with regularity.[3]

The emphases in the literature were almost exclusively on connections with the West; that Russia might have been purely a recipient of influence (*vliianie*) by French, German, or English

---

[2] The deleterious impact of Pavel Berkov, who long headed the Group for the Study of Eighteenth-Century Russian Literature at the Institute of Russian Literature [Pushkin House], is emphasized by David M. Griffiths, "In Search of Enlightenment: Recent Soviet Interpretation of Eighteenth-Century Russian Intellectual History," *Canadian-American Slavic Studies* 16, nos. 3-4 (Fall-Winter 1982): 317-56. For a pungent explication of the topic, which argues that the "Russian Enlightenment" was largely the product of baneful posturing by select Soviet and East German scholars, see Max J. Okenfuss, *The Rise and Fall of Latin Humanism in Early-Modern Russia: Pagan Authors, Ukrainians, and the Resiliency of Muscovy* (Leiden: E. J. Brill, 1995), 223-30.

[3] An eminently useful reference is the series *XVIII vek* (Moscow-Leningrad-St. Petersburg, 1935-2011). Comprising twenty-six volumes thus far, it is, despite continuing excessive claims by some of its contributors concerning the efflorescence and originality of eighteenth-century Russian culture, an excellent survey of Russian intellectual life and, it should be said, of the Russian Enlightenment.

cultural forces without some presumed reciprocal Russian impact on "European culture" was, at least formally, long rejected. If Russia underwent Enlightenment, then the requisite presence of *prosvetiteli* (enlighteners) is obvious.[4] Lomonosov, the "first Russian scientist," was a clear candidate for canonization as the *"velikii syn russkogo naroda"* ("great son of the Russian people").

After all, his lowly, non-noble, background harmonized marvelously with the quasi-Marxist tenets that many Soviet historians and literary specialists were forced to pay obeisance to in their studies of the Russian eighteenth century. Equally impressed, however, by the seemingly stark contrast between Lomonosov's plebian upbringing and his attainments were eighteenth- and nineteenth-century writers who, for reasons and objectives that will be considered, made his childhood struggles to surmount all manner of social and economic impediments central to their reverent accounts of his life.

As was often typical for a natural philosopher in the eighteenth century, the scope of Lomonosov's interests and activities was protean. Aside from dissertations in chemistry, physics, metallurgy, mining, geology, astronomy, and on the administration of science in Russia,[5] he composed several literary and linguistic treatises, including a manual on rhetoric, a Russian grammar, and a proposed series of reforms for Russian versification. Lomonosov is also remembered for being one of Russia's most notable poets,

---

[4]   As noted by Griffiths, "In Search of Enlightenment," 317.

[5]   The majority of Lomonosov's writings were in natural philosophy, widely defined. See the latest and arguably definitive version of his collected works: M. V. Lomonosov, *Polnoe sobranie sochinenii* (*PSS*), vols. 1-11 (Moscow-Leningrad, 1950-83), especially vols. 1-5; in addition vols. 9-11 contain extensive official documentation and correspondence related to his scientific work. The notes to individual papers in the series disclose previous publication data. G. Z. Kuntsevich, comp. *Bibliografiia izdanii sochinenii M. V. Lomonosova na russkom iazyke* (Petrograd, 1918), charts the issuance of several earlier editions of Lomonosov's collected works. For Lomonosov's eighteenth-century Russian language publications, see *Svodnyi katalog russkoi knigi grazhdanskoi pechati XVIII veka, 1725-1800*, 6 vols (Moscow, 1963-75), particularly volume 2, 162-77.

a less remarkable dramatist, and the author of once widely-disseminated historical works. For a time he directed the Academy of Sciences' gymnasium and university, oversaw its geographical department, helped supervise the Academy's publishing activities, founded Russia's first chemical laboratory, assisted in establishing Moscow University, opened a factory devoted to glass production, expended enormous energy in developing the mosaic arts in Russia, and worked on devising scientific instruments, perhaps most conspicuously those meant to aid Russian navigational endeavors.

Lomonosov has been uniformly extolled within Russia as one whose contributions to *nauka* (science),[6] undeservedly neglected though they might be outside of Russia, do not pale in comparison with those of such scientific, cultural, and ultimately national icons as Newton, Copernicus, Galileo, and Benjamin Franklin. Analogies to Newton and Franklin especially are inscribed in the historiography on Lomonosov, and tellingly underscore the lofty stature assigned to him in Russian cultural discourse. But unlike in the cases of the above "worthies of science," there are no sure discoveries or paradigm-shattering insights universally attributed to him. Russian scholars have taken great pains to correct this apparent deficiency, and their efforts to broadly inculcate the notion that Lomonosov's fertile scientific speculations demonstrate profound originality and prescience have proceeded at an escalating pace over the past two centuries.[7]

---

6   *Nauka*, frequently translated as science, has a broader meaning than its English equivalent does and is better compared to the German *Wissenschaft*, which connotes a diffuse pursuit of knowledge not confined to natural philosophy. The distinction between science and the more expansive *nauka* will become evident in the ensuing analysis. Referring to Lomonosov or any early modern chemist, physicist, astronomer, mathematician, etc., as a scientist is, of course, anachronistic (the term itself was not widely used until the early decades of the nineteenth century), but it has become a conventional marker that eases semantic confusion.

7   The "Lomonosov industry" has been a fantastically prolific one: by my count it includes some four thousand publications, and the number continues to grow. The leading part played by the Academy of Science in organizing this devotional effort is covered in M. I. Radovskii, *M. V. Lomonosov i Peterburgskaia Akademiia nauk* (Moscow-Leningrad, 1961), 222-71. G. I. Sma-

A commonplace in the historical literature on Russian science is the ostensibly concomitant assumption that Lomonosov's researches in chemistry, physics, geography, and whatever else his many and varied work habits led him to can be concretely linked to the work of successive generations of scientists. The highly speculative nature of Lomonosov's scientific papers, in addition to

---

gina (*Kniaginia i uchenyi: E. P. Dashkova i M. V. Lomonosov* [St. Petersburg, 2011]) examines the Academy's efforts in the 1780s and 1790s to both publish Lomonosov's writings and to encourage biographical writings about him. These endeavors were overseen by Princess Ekaterina Dashkova, the Academy's most energetic eighteenth-century director. For a guide to most of the pre-Soviet literature, which makes up less than a quarter of the total, see A. G. Fomin et al., eds., *Materialy po bibliografii o Lomonosove na russkom, nemetskom, frantsuzskom, ital'ianskom i shvedskom iazykakh* (Petrograd, 1915). For more recent sources see the bibliographical and/or archival materials contained in each volume of *Lomonosov: sbornik statei i materialov* (Moscow-Leningrad-St. Petersburg, 1940-2011). Nearly all pertinent archival information concerning Lomonosov's own writings can be found in L. B. Modzalevskii, ed., *Rukopisi Lomonsova v Akademii nauk SSSR: nauchnoe opisanie*, with a preface by B. N. Menshutkin (Moscow-Leningrad, 1937); L. B. Modzaelvskii, I. V. Tunkina, eds., *M. V. Lomonosov i ego literaturnye otnosheniia v Akademii nauk: Iz istorii russkoi literatury i prosveshcheniia serediny XVIII v.* (St. Petersburg, 2011); E. S. Kuliabko and E. B. Beshenkovskii, *Sud'ba biblioteki i arkhiva M. V. Lomonosova* (Leningrad, 1975); I. M. Beliaeva, ed., *Biblioteka M. V. Lomonosova: nauchnoe opisane rukopisie i pechatnykh knig* (Moscow, 2010); and G. G. Martynov, ed., *Mikhail Vasil'evich Lomonosov: perepiska, 1737-1765* (Moscow, 2010). Moreover, the extensive commentaries to Lomonosov's collected works located at the conclusion of each volume, dispense abundant references to germane primary and secondary literature. Lomonosov's science received comparatively less attention than his belletristic side until the end of the nineteenth century. Since then there has a rough parity in space allotted to his scientific and literary activities. Soviet historians of science were unmatched, in quantitative terms, in issuing biographies (and to a far lesser extent autobiographies) of scientists, natural philosophers, technical specialists, and the like. Herculean biographical efforts to exhibit chiefly Russian scientific progress effectively comprised the entirety of Soviet studies of the scientific past. These studies are ignored in virtually all non-Russian language scholarship on the "rise of European science." An arresting historiographical example is Thomas Söderqvist, ed., *The History and Poetics of Scientific Biography* (Burlington, VT: Ashgate, 2007); see particularly his introduction. Söderqvist's commendable aim of surveying the scale of scientific biography since its inception is undercut by omitting, except for an incorrect allusion, references to Russian-language scholarship.

the unfinished state in which he left many of them, allowed scholars working in the shadow of the expansive renown his name achieved after his death to engage in extraordinary inferences in regard to his apparent connection to later scientists, along with their discoveries and conjectures.[8]

Although attempting to delineate direct intellectual influence is fraught with pitfalls, avowals such as that of the historian Mikhail Sukhomlinov that: "Rumovskii, Kotel'nikov, and Protasov received their scientific education under Lomonosov; Lepekhin and Inokhodstev were the students of Rumovskii and Kotel'nikov; Ozeretskovskii, Sokolov and Severgin had their views formed under the beneficial influence of Lepekhin etc.,"[9] have exerted a tenacious hold on Russian and Soviet scholars evaluating Lomonosov's place in the history of science. While the aforementioned natural philosophers, active in the late eighteenth and early nineteenth centuries, were all certainly aware of Lomonosov's scientific work, and several of them knew him personally, there is no evidence of a lineage leading from Lomonosov's scientific treatises to the substance of their respective studies. This is true of his eighteenth-century contemporaries, and markedly true of any presumed line of descent, uninterrupted or not, between Lomonosov and later generations of scientists.

In the more easily delimited area of whether Lomonosov created a school or community of students who carried on his work in the sciences, it can be categorically acknowledged that he left none. The only pupil trained by Lomonosov who unmistakably attempted to follow in his footsteps, Vasilii Klement'ev, served as

---

8   Eventual archetypes were Boris Menshutkin's *Lomonosov kak fiziko-khimik: k istorii khimii v Rossii* (St. Petersburg, 1904); and idem, *Mikhailo Vasil'evich Lomonosov: zhizneopisanie* (St. Petersburg, 1911).

9   M. I. Sukhomlinov, *Istoriia Rossiiskoi Akademii*, vol. 4 (St. Petersburg, 1878), 2. Stepan Rumovskii (mathematician), Semen Kotel'nikov (mathematician), Aleksei Protasov (anatomist), Ivan Lepekhin (explorer), Petr Inokhodtsev (astronomer), Nikolai Ozeretskovskii (naturalist), Nikolai Sokolov (chemist), and Vasilii Severgin (chemist and mineralogist) were among the most illustrious figures of early Russian science.

his assistant in chemistry, but predeceased him by more than five years (Klement'ev died in 1759).[10] Moreover, Lomonosov had largely abandoned active work in his chemical laboratory and the training of students by the early 1750s. Despite the assertions of many Russian and Soviet scholars, such esteemed eighteenth-century natural philosophers as Rumovskii and Kotel'nikov assiduously avoided Lomonosov's embrace. Rumovskii in particular, as will be seen, was scathing in his view of Lomonosov's scientific abilities, and can hardly be classified as a follower of his.

However, rather than dwelling on or excessively contesting the well-trodden minutiae of Lomonosov's biography, at least beyond what is necessary to grasp the contours of his impact on Russian culture, here we will focus on an attempt to understand why a mythology of Lomonosov took shape, and what its evolving significance is. It is indisputable that an exaggeratedly rich intellectual genealogy in Russian science, with Lomonosov cast as the progenitor of a host of nascent scientific disciplines and advancements, has existed since the late nineteenth century at the latest.

Foundational elements for this mythology are, however, already encountered in memoirs of Lomonosov written in the last three decades of the eighteenth century. The highly selective

---

[10]   Lomonosov, *PSS*, vol. 9, 60-63, 103, 442-43, 471-72, 664, 667-68, 675-79, 852; N. M. Raskin, *Khimicheskaia laboratoriia M. V. Lomonosova* (Moscow-Leningrad, 1962), 130-40; and idem, *Vasilii Ivanovich Klement'ev — uchenik i laborant M. V. Lomonosova* (Moscow-Leningrad, 1952). Nathan Brooks blames Lomonosov's failure to train any successors on the absence of a stable community of scientists in eighteenth-century Russia. There were not, in his view, any established institutional processes by which students could succeed their teachers. See Nathan Marc Brooks, "The Formation of a Community of Chemists in Russia: 1700-1870" (Ph.D. diss., Columbia University, 1989), 40-58. Brooks's scientific communities' thesis is unobjectionable, if also overly narrow; future studies of the structures of science in "early modern" Russia might benefit from investigating the nature of both formal and informal patron-client networks. A thought-provoking work of this type, focusing on Galileo's tactics and strategies for advancement at, principally, the Florentine court, is Mario Biagiolo, *Galileo, Courtier: The Practice of Science in the Culture of Absolutism* (Chicago: University of Chicago Press, 1993).

configuration of historical details in these accounts testifies strongly to certain "mythogenic" qualities in Russian culture that seem to have been crucial in not only structuring the content of these memoirs, but in decisively determining their reception.[11] Lomonosov's autobiographical reflections have also been a critical resource for later representation and distortion.

While the elevation of scientists to secular sainthood, with the accompanying inaccuracies, exaggerations, or falsehoods that mark their received biographical lives, is hardly unique to Russian culture, there are singularities that characterize the birth of any myth.[12] The mythmaking temper of eighteenth-century Russia,

---

[11]   Irina Reyfman offers an instructive analysis of this phenomenon, and more specifically of the formation of eighteenth-century Russian literary mythologies and Lomonosov's preeminent position in them in her *Vasilii Trediakovsky: The Fool of the 'New' Russian Literature* (Stanford: Stanford University Press, 1990), 1-131. Underlining the power of the creation myth in eighteenth-century Russia, she notes, apropos of the role of figures such as Peter the Great and Lomonosov: "The main character in a creation myth, a demiurge or cultural hero, gives things their proper disposition and sets rules for future generations.... The hero is thus in a sense the ancestor of the present community" (ibid., 11). For more on the mythological ethos that seem to have distinguished eighteenth-century Russia, see the following seminal articles: Iu. M. Lotman, "The Poetics of Everyday Behavior in Russian Eighteenth-Century Culture," in Iu. M. Lotman and B. A. Uspenskii, *The Semiotics of Russian Culture*, ed. Ann Shukman, trans. N. F. C. Owen (Ann Arbor, MI: Department of Slavic Languages and Literatures, University of Michigan, 1984), 231-56; Iu. M. Lotman and B. A. Uspenskii, "The Role of Dual Models in the Dynamics of Russian Culture (Up to the End of the Eighteenth Century)," in ibid., 3-35; and idem, "Myth-Name-Culture," in *Soviet Semiotics: An Anthology*, ed. and trans. Daniel P. Lucid (Baltimore: John Hopkins University Press, 1988), 242-43.

[12]   See Pnina G. Abir-Am and Clark A. Elliot, eds., *Commemorative Practices in Science: Historical Perspectives on the Politics of Collective Memory* (published in *Osiris* 14; Ithaca, N. Y., 1999), for a discussion of the diverse purposes by which national-political, institutional, and disciplinary agendas might be satisfied or thwarted by manipulating the more visible imagery devoted to select scientific "cultural heroes" (the chapters dealing with Copernicus, Louis Pasteur, and Max Plank are especially interesting). Although Franklin's science is not its focus, of comparative value to my work is Nian-Sheng Huang, *Benjamin Franklin in American Thought and Culture, 1790-1990* (Philadelphia: American Philosophical Society, 1994). Profitable also is

which enabled Lomonosov's reputation to develop to astounding proportions, seems to have derived its strength from the more momentous, indeed quite omnipotent, historical presence of Peter the Great.[13] His reign was long invested with consummately apocalyptic meanings by many Russians.

Central to conceptions of the Petrine epoch was the idea that *staraia Rossiia* (the old Russia), and its attendant culture, had been thoroughly vanquished by *novaia Rossiia* (the new Russia). This had the result, according to Iurii Lotman and Boris Uspenskii, that "the 'new' was identified with all that was good, valuable and worthy of emulation," "while "the 'old' was thought to be bad, due for

---

François Azouvi, "Descartes," in *Realms of Memory: The Construction of the French Past*, vol. 3: *Symbols*, ed. Pierre Nora, trans. Arthur Goldhammer (New York: Columbia University Press, 1998), 483-521. Azouvi traces the path of Descartes's reputation in France over the past three centuries. As averred by Azouvi, Descartes and Cartesianism have been subject to such intensely competing, and obscuring, political, religious, and scholarly pressures by successive generations of French writers that it is difficult to speak precisely of what constitutes either Descartes's biography or Cartesian philosophy. As for the mythology surrounding Newton's life, a similarity between the methods and aims of his early biographers and Lomonosov's memoirists is suggested in later chapters of this volume.

[13] On the origins and unfolding of the cult of Peter the Great in Russia, consult the following: D. K. Burlaka et al., eds. *Petr Velikii--pro et contra: lichnost' i deianiia Petra I v otsenke russkikh myslitelei i issledovatelei: antologiia* (St. Petersburg, 2003); Michael Cherniavsky, *Tsar & People: Studies in Russian Myths*, 2nd ed. (New York: Random House, 1969), 72-100; Xenia Gasiorowska, *The Image of Peter the Great in Russian Fiction* (Madison, WI: University of Wisconsin Press, 1979); Lindsey Hughes, *Peter the Great: A Biography* (New Haven: Yale University Press, 2002), 226-50; Iu. M. Lotman, "Echoes of the Notion of 'Moscow as the Third Rome'" in Peter the Great's Ideology," in Shukman, *The Semiotics of Russian Culture*, 53-67; S. I. Nikolaev, ed. *Peter I v russkoi literature XVIII veke: teksty i kommentarii* (St. Petersburg, 2007); Kevin M. F. Platt, *Terror and Greatness: Ivan and Peter as Russian Myths* (Ithaca, NY: Cornell University Press, 2011); M. Pliukhanova, "'Istoricheskoe' i 'mifologicheskoe' v rannykh biografiakh Petra I," in *Vtorichnye modeliriushchie sistemy* (Tartu, 1979), 82-88; B. N. Putilov, ed., *Petr Velikii v predaniiakh, legendakh, anekdotakh, skazkakh, pesniakh* (St. Petersburg, 2000); Nicholas V. Riasanovsky, *The Image of Peter the Great in Russian History and Thought* (New York: Oxford University Press, 1985); and E. Shmurlo, *Petr Velikii v otsenke sovremennikov i potomstva* (St. Petersburg, 1912).

destruction and demolition."[14] From this belief was generated the resolute conviction, widespread among elites, that commencing with the era of Peter the Great Russians had experienced not merely a cultural reawakening but nothing less than an entirely "new beginning" that had reoriented their very thinking.[15]

Certainly the latent, and hence disturbing, potentialities of science and the scientist were pivotal to the reasons Peter's rule was perceived as such a transformative break with tradition. Lomonosov, motivated by a selfless desire to further learning among his countrymen, personified the ideals of the Petrine era. He served initially as the vehicle that induced acceptance of this new type of knowledge, and later as its primary propagator. Both in his personal qualities and in his professional attainments, Lomonosov's

---

[14]   Lotman and Uspenskii, "The Role of Dual Models," 18. This "image of 'the new Russia' and 'the new people' became a special kind of myth which came into existence already at the beginning of the eighteenth century and was passed on to the later cultural consciousness," and which, assert Lotman and Uspenskii, "has become so deeply rooted that it has in fact never seriously been questioned." See also Iu. V. Stennik, *Ideia "drevnei" i "novoi" Rossii v literature i obshchestvenno-istoricheskoi mysli XVIII - nachala XIX veka* (St. Petersburg, 2004); and Joachim Klein, *Puti kul'turnogo importa: trudy po russkoi literature XVIII veka* (Moscow, 2005), especially his chapter: "Rannee Prosveshchenie, religiia i tserkov' y Lomonosova."

[15]   Stephen L. Baehr, *The Paradise Myth in Eighteenth-Century Russia: Utopian Patterns in Early Secular Russian Literature and Culture* (Stanford: Stanford University Press, 1991), offers a wide-ranging examination of eighteenth-century literature on Peter the Great; the published works on the topic were, not unexpectedly, wholly panegyric in tone. The place of Peter in Lomonosov's writings, perhaps witnessed with particular clarity through the medium of his laudatory odes to Peter's daughter Elizabeth, is covered in many studies, including the aforementioned by Baehr. See also V. P. Grebeniuk, "Petr I v tvorchestve M. V. Lomonosova, ego sovremennikov, predshestvennikov i posledovatelei," in A. S. Kurilov, ed., *Lomonosov i russkaia literatura* (Moscow, 1987), 64-80; Marcus C. Levitt, *The Visual Dominant in Eighteenth-Century Russia* (DeKalb, IL: Northern Illinois University Press, 2012), 15-63; Elena Pogosian, *Vostorg russkoi ody i reshenie temy poeta v russkom panegirike 1730-1762* (Tartu, 1997), 85-123; and Il'ya Z. Serman, *Mikhail Lomonosov: Life and Poetry*, trans. Stephany Hoffman (Jerusalem: Hebrew University, 1988), 82-112. Nicholas Riasanovsky has tracked the central place that Peter I occupies in Russian historical discourse. Lomonosov's views of the "Tsar-Reformer" are expertly presented (see *Image of Peter the Great*, 30-34, 50).

biography signified an individual of superhuman (indeed of Petrine) dimensions. The eventual conflation of his life with both the myth of Peter the Great, albeit in a distinctly supporting role,[16] and with the complementary notion of a revolutionary pace of change that seemingly characterized the entire eighteenth century, broadly reveals the genesis of what he came to mean historically. Lomonosov incarnated the Petrine (and Soviet?) ideal of rank achieved through meritorious service contrasted against the allegedly discredited acquisition of rank by birth alone.

This volume traces the origins and development of a prodigious imagery devoted to representing Lomonosov as the father of Russian science from its forging in the late eighteenth century to its demise at the end of the Soviet experiment.[17] Idealized depictions of Lomonosov were employed by Russian scientists, historians, and poets, among others, in efforts to demonstrate to their countrymen and to the State the pragmatic advantages of

[16]   An article by Aleksandr Portnov, "Nu, Mikhailo Vasilich, zadal zagadku. Byl li Lomonosov vnebrachnym synom Petra I?," *Trud* 65 (13 April 1995), brings up a tale that holds Lomonosov to have been Peter's illegitimate son. This piece is cited by Lindsey Hughes in *Russia in the Age of Peter the Great* (New Haven, CT: Yale University Press, 1998), 331. Hughes, of course, dismisses the "legend" that Lomonosov was Peter's issue, as does Portnov, while also noting that Lomonosov "was undoubtedly Peter's spiritual offspring."

[17]   In the epilogue I speculate on the "afterlife" or post-Soviet life of the myth. I have restricted my investigation to written representations of Lomonosov. Although in this type of inquiry the author is always subject to charges of idiosyncrasy in their choices of texts, I believe that the selection of specific writings and authors becomes sufficiently clear in the presentation. There also exists an array of visual imagery honoring Lomonosov, much of it strikingly hagiographic. For these non-textual portrayals, see D. S. Babkin, "Obraz Lomonosova v portretakh XVIII v.," in *Lomonosov: sbornik statei i materialov*, vol. 1 (Moscow-Leningrad, 1940), 302-17; V. L. Chenakal, *M. V. Lomonosov v portretakh, illiustratsiakh, dokumentakh* (Moscow-Leningrad, 1965); M. E. Glinka, *M. V. Lomonosov (opyt ikonografii)* (Moscow-Leningrad, 1961); *Lomonosov v knizhnoi kul'ture Rossii* (Moscow, 2010); S. I. Nikolaev, "Ranniaia ikonografiia Lomonosva v svete ikonologii," in *XVIII vek* 26 (2011): 73-84; and V. V. Rytikova, "Obraz M. V. Lomonosova v monumentalynikh zamyslakh Leningradskikh skul'ptorov 1960-1980-kh gg.," in *Lomonosov: sbornik statei*, vol. 10 (St. Petersburg, 2011), 325-42.

science to a modernizing nation. The idea that science was critical to the fulfillment of wider cultural aspirations was also embedded in his deification. I have detached Lomonosov's scientific legacy from perceptions of his significance as a litterateur. Although he himself may have viewed chemistry and physics as his main occupations, the legacies of national heroes are contested terrain, and his life has been utilized by later writers to further a variety of scholarly and historical agendas.[18] In setting forth this assumption, I maintain that no sharply drawn division can be upheld between the utilization of the myth of Lomonosov during the Soviet period of Russian history and that which characterized earlier views. The main elements that formed the mythology were laid down in the eighteenth and nineteenth centuries; Soviet scholars simply added more exaggerated layers to existing representations.

---

[18]   V. P. Zubov, *Istoriografiia estestvennykh nauk v Rossii* (*XVIII v. – pervaia polovina XIX v*) (Moscow, 1956). Zubov's historiographical review has not only retained its value for the study of, in particular, eighteenth-century Russian science, but provides a fairly thorough survey of writings on Lomonosov up to the mid-nineteenth century. A less reliable foray into the literature on Lomonosov is Iu. I. Solov'ev and N. N. Ushakova, *Otrazhenie estestvennonauchnykh trudov M. V. Lomonosova v russkoi literature XVIII i XIX vv* (Moscow, 1961). The authors are rather too determined to illustrate Lomonosov's sway over later Russian thought, scientific and otherwise. For a more careful historiographical examination, inclusive of both older and more recent scholarship, focusing on the history of Russian chemistry, see Z. I. Sheptunova, *Istoriogrograficheskii analiz rabot po istorii khimii v Rossii XVIII—nachalo XX v* (Moscow, 1995). Promoted as the first "post-Soviet" biography of Lomonosov, Valerii Shubinskii's worthy *Mikhail Lomonosov: vserossiiskii chelovek* (St. Petersburg, 2006) questions many of the hagiographical excesses that distort Lomonosov's place in Russian history and in the history of science; Shubinskii then does little to overturn them in a very readable account. Perhaps the best of the "western" studies that treat Lomonosov's scientific life are: Valentin Boss, *Newton and Russia: The Early Influence, 1698-1796* (Cambridge, MA: Harvard University Press, 1972), 152-237; *Mikhail Vasil'evich Lomonosov on the Corpuscular Theory*, ed. and with an introduction by Henry M. Leicester (Cambridge, MA: Harvard University Press, 1970), 3-48; and Alexander Vucinich, *Science in Russian Culture: A History to 1860* (Stanford: Stanford University Press, 1963), 105-16, 401-02.

A scientist's activity in heterogeneous areas, though not at all unusual in the eighteenth century, became, like that of the very idea of an encyclopedic figure, incompatible with the unfolding of narrower professional specializations in the nineteenth century. Commencing in the late nineteenth and early twentieth centuries, students intent on defining Lomonosov's place in Russian science concentrated on dividing his hitherto myriad roles into those of either a chemist, a physicist, or a geographer, etc.[19] Although the details of Lomonosov's scientific labors were creatively broadened in subsequent retellings, representations of him as the embodiment of the arrival and rise of science in Russia were the fulcrum on which nearly all accounts were built.

Lomonosov's ardent efforts to fashion a secure "socioprofessional" role for himself as a natural philosopher at the Academy of Sciences are addressed in the first chapter. A scientific vocation was as yet an ill-formed occupational category, quite lacking in established rank. Those who successfully pursued scientific careers were utterly dependent on the favor of powerful benefactors. When he was seeking tangible support, or merely encouragement, Ivan Shuvalov was Lomonosov's most reliable patron. Lomonosov's rather adept use of patronage to advance his status in Russian society shaped his own mythopoetic endeavors. His related attempts to closely associate himself with the prestige of Christian Wolff and Leonhard Euler are scrutinized in this chapter. His apparent links with Wolff and Euler are exceptionally important motifs, first in his

---

[19]   For, as noted by John Gascoigne: "Science no less than religion needs its gallery of saints as sources of emulation to provide a sense of continuity and tradition. But, inevitably, posterity is selective in drawing up such a roll-call of the blessed as the past is scavenged for figures that seem best to conform to the needs of the present. Scientists of the nineteenth and twentieth centuries accorded most respect to the founding fathers of their discipline that left their mark in the manner most familiar to scientists of a later age." John Gascoigne, "The Scientist as Patron and Patriotic Symbol: the Changing Reputation of Sir Joseph Banks," in *Telling Lives in Science: Essays on Scientific Biography*, ed. Michael Shortland and Richard Yeo (Cambridge: Cambridge University Press, 1996), 243.

own expressed self-perceptions and later in historical representations of him.

There are manifest indications that numerous Russian thinkers interested in emphasizing the importance of natural philosophy in their country's development were deeply inspired, if not overtly influenced, by the heroic image, or mythology, of Lomonosov that had become pervasive by the end of the eighteenth century. The development of Lomonosov's scientific biography by several of his "contemporaries" is the subject of Chapter 2. Writings by Jacob von Staehlin, Nikolai Novikov, and Mikhail Verevkin were fundamental in shaping early views of Lomonosov. Subsequent scholars constantly revisited their evaluations of Lomonosov, as well as those authored by Mikhail Murav'ev and Aleksandr Radishchev. Radishchev's incisive appraisal of Lomonosov has had an especially interesting, if ambivalent, resonance in the mythology. This chapter illustrates that the image of Lomonosov served as an avenue through which trends in scientific thought were discussed and to an extent "popularized" in eighteenth-century Russia. These early memoirs of Lomonosov also implicitly highlight the embryonic growth of the biographical genre in Russia.

Chapter 3 explores attempts to define Lomonosov's worth as a natural philosopher, which were undertaken at the beginning of the nineteenth century by the chemist and mineralogist Vasilii Severgin. His insistence on the continuing relevance of Lomonosov as the worthiest of exemplars for future Russian scientists, with a specific emphasis on Lomonosov's work in establishing the importance of science to the Russians, highlights the continuing search for status among early Russian scientists. That Severgin was mainly a "professional" scientist made his evaluation a notable resource. Alexander Pushkin's eloquent assessments of Lomonosov's overall place in Russian culture proved, owing to Pushkin's totemic status in Russian life, to be of great import in further developing Lomonosov's historical prominence and are also discussed in this chapter. Examining associations between Pushkin and Lomonosov provides essential further insights into the strength of a mythological ethos in Russian culture.

A fascinating encounter between nineteenth-century Russian academics and writers and the extensive "scientific" imagery that had accrued to Lomonosov's name occurred during the 1855 Moscow University centennial celebrations. While Lomonosov's cultural achievements received wide acclaim, his science was also subjected to the kind of searching critique to which it had never before been exposed. This new response to Lomonosov's reputation is analyzed in Chapter 4. Touched on as well is the literature that emerged in conjunction with the 1865 Lomonosov Jubilee commemorations. Gatherings were held in more than twenty cities and towns. The many publications on Lomonosov issued that year crystallized efforts by cultural figures of all hues to implant the idea that Russia was becoming an increasingly modern nation characterized by an established scientific heritage. That Lomonosov exemplified the spirit of this inheritance was made explicit both at the Moscow University proceedings in 1855 and throughout the jubilee of 1865.

The chemist and historian of science Boris Menshutkin steadfastly devoted himself to enhancing both qualitatively and quantitatively the historiography of Lomonosov's science. Menshutkin's nearly four decades of labor in Russian and Soviet archives (which ended with his death in 1938) would serve as the basis for the over twenty studies on Lomonosov he published, and are considered in the concluding chapter. In bringing to light Lomonosov's previously unpublished or seemingly forgotten chemical and physical manuscripts, along with adding extensive commentaries to them, Menshutkin strove to superimpose an extensive scholarly apparatus onto the already impressive scientific legacy ascribed to Lomonosov. His persistent emphasis on the anticipatory nature of Lomonosov's scientific speculations permeates all later works and is the apotheosis of Lomonosov's image as an intrepid scientific discoverer.

It was especially through the framework of a popular biography of Lomonosov that Menshutkin first issued in 1911 that the combination of analysis and legend in the Lomonosov myth was most decisively attained. Menshutkin's role in elaborating upon the mythology crosses the somewhat artificial historical divide between Imperial Russia and the Soviet Union, and fittingly demonstrates that representations of Lomonosov as the first Russian scientist were not solely the product of any specific political posture, but rather appeared and were kept relevant due to the efforts of generations of Russian thinkers.

Menshutkin concluded an expanded edition of this biography by quoting from Lomonosov's translation of Horace's *Exegi monumentum*:

> I have reared myself a monument of immortality
> Higher than the pyramids, and stronger than brass,
> Stormy Aquilon cannot break it,
> And it will not be overwhelmed by the passage of centuries.
> I shall never wholly die, and death will leave aside
> The greatest part of me, when my life is at an end.[20]

Horace's ode (it was later more famously rendered into Russian by Pushkin), beautifully allegorizes not only Lomonosov's apparently successful quest for earthly honors, but also, as is clear, his desire to be memorialized by succeeding generations of his compatriots. The great praise with which Russian culture has long endowed Lomonosov's name suggests that his goals were achieved.

As is the case with Pushkin, Lomonosov's fame has far surpassed any realistic association with the known details of his biography; Lomonosov's monument is the mythology: how and why it was created is more intriguing than his actual scientific accomplishments.[21] Indeed, Lomonosov is, I would argue, of interest

---

[20]   Menshutkin, *Zhizneopisanie Mikhaila Vasil'evicha Lomonosova*, 2nd ed. (Moscow-Leningrad, 1937), 236. Lomonosov's version of Horace is from his *Brief Guide to Eloquence* (*Kratkoe rukovodstvo k krasnorechiiu*, 1748). See Lomonosov, *PSS*, vol. 7, 314.

[21]   Pierre Nora's attempt to interpret the French past in way that is "less interested in causes than in effects; less interested in actions remembered or even commemorated than in the traces left by those actions and in the interaction of those commemorations; less interested in the events themselves than in the construction of events over time, in the disappearance and reemergence of their significations; less interested in 'what actually happened' than in its perpetual reuse and misuse, its influence on successive presents; less interested in traditions than in the way in which traditions are constituted and passed on," can be used fruitfully in the consideration of certain Russian myths and symbols. See preface to Pierre Nora, ed. *Realms*

primarily as a symbolic figure, until recently an extraordinarily resilient one, who over the course of two centuries came to fulfill the tangible intellectual and emotional requirements that Russian pride demanded in a national myth.[22]

---

of Memory: Rethinking the French Past, vol. 1: Conflicts and Divisions, trans. Arthur Goldhammer (New York: Columbia University Press, 1996), XXIV.

[22]   For the origins of Russian pride, or national identity, or national consciousness, or nationalism—as with most of the vast literature on these topics, the definitions are hopelessly blurred when applied to specific "national" conditions—see Liah Greenfeld's controversial, always challenging, explanation: Liah Greenfeld, Nationalism: Five Roads to Modernity (Cambridge, MA: Harvard University Press, 1992), 189-274. For Greenfeld, "in the final analysis, it was 'native Russianness' that justified the new status (or status aspirations) of the non-noble intellectuals" (such as Mikhail Lomonosov, p. 243), and this incipient Russianness developed exclusively out of "ressentiment" of the "West." The "existential" resentment, jealousy, felt by the Russian elite, both noble and non-noble, towards England, France, Prussia, the Netherlands, and other countries, as Greenfeld describes, was also where Lomonosov's "bitter and unjustified" hatred of other scientists, especially Germans at the Academy of Sciences, found its source. Her arguments in part illuminate why and how Lomonosov himself became such a symbol for national (or nationalist) aspirations: "he symbolized the need to aggrandize Russian culture and make it comparable to the cultures of Western Europe."

Chapter 1

Honor and Status in Lomonosov's "Autobiography"

*M*ikhail Lomonosov's reputation as a natural philosopher grew dramatically in the years immediately following his death. This fact does suggest that prior to this posthumous devotion, which took shape most distinctly through a surfeit of biographical encomiums, Lomonosov's name was in danger of falling into obscurity in Russia. The exact mechanisms whereby great renown originally became attached to his life, however, are unclear.[1] The quite discernible mythogenic features in eighteenth-century Russian culture partially explain the development, but it is Lomonosov's zealous and skillful advocacy of his own image that is especially interesting. This aspect of the creation of his biography in many senses still determines how aspects of his life are perceived, and it will be defined, with deference to Stephen Greenblatt, as Lomonosov's self-fashioning.

Greenblatt asserts persuasively in his studies that the capacity to shape one's self and one's autonomy was, by the sixteenth century, progressively constrained by the capability of "family,

---

[1] Richard Yeo makes a similar point about the "precise origin and development of the elements that constitute the Newtonian mythology" (Richard Yeo, "Genius, Method, and Morality: Images of Newton in Britain, 1760-1860," *Science in Context* 2, no. 2 [Autumn 1988]: 258-59). Despite uncertainty regarding its inception, his subsequent emphasis on the enveloping ubiquity of that mythology throughout eighteenth-century England is convincingly presented.

"Lomonosov"
(lightning is visible in the corner of the window; a symbol of the
scientist's study of atmospheric electricity)
Engraving by E. Fessar, 1757

state, and religious institutions" to enforce "discipline" on a realm's subjects (at first mainly the elite). Self-fashioning therefore scarcely denotes the power of an individual to "govern the generation of identities" altogether; rather, the act of affecting an identity entailed "submission to an outside power or authority situated at least partially outside the self—God, a sacred book, and institutions such as church, court, colonial or military administration."[2] Transposing Greenblatt's claims—that literary life in Renaissance England was marked by an "increased self-consciousness about the fashioning of human identity as a manipulable, artful process," if also a process increasingly circumscribed by the limiting power of a variety of societal "structures"—to eighteenth-century Russia,[3] it is evident that the paths pursued by Lomonosov resemble nothing so much as the advancement strategies adopted by a "profoundly mobile," educated, outsider. This would be a person who, lacking high social status and desirous of succeeding in a hierarchical society, always seeks association with and the protection of powerful figures close to the locus of authority, the court.

Fashioning a life, "real" or fictional, is a protracted "process of negotiation" whereby the self is never static; instead it approximates the varieties of selves made available by society to the subject at a specific time and place.[4] Contingent on the claims made upon him, Lomonosov was able, rather adeptly, to simultaneously to enact the role of natural philosopher, chemist, poet, historian, academician, educator, geographer, rhetorician, and artist, while ceaselessly rearranging the elements that constituted his life in order to publicly represent that life. Two centuries of biographers would continue this process of negotiation. As this work will illustrate, however, political, social, and cultural forces predominating first in Imperial

---

2   Stephen Greenblatt, *Renaissance Self-Fashioning: From More to Shakespeare* (Chicago: University of Chicago Press, 1980), 1, 9.

3   Ibid., 1-2.

4   David Aubin and Charlotte Bigg, "Neither Genius nor Context Incarnate: Norman Lockyear, Jules Janssen and the Astrophysical Self," in Söderqvist, *History and Poetics*, 65.

Russia and then in the Soviet Union would increasingly restrict, qualitatively if not quantitatively, the array of features that would be allowed to adorn the written story of Lomonosov's life.

Mario Biagioli cautiously employed Greenblatt's idea(s) in his study of Galileo's shaping of his "socioprofessional" persona as both a philosopher and a mathematician, or rather as a "philosophical astronomer," a decidedly new and fragile combination, at the court of the Medicis and at the Vatican.[5] Although Biagioli concentrates on revealing Galileo's exploitation of patronage, he also demonstrates that Galileo's mimetic strategies effectively constructed his public image—then and for posterity.

Lomonosov's use of patronage is laid bare in this analysis of the origins of the heroic imagery that surrounded him, for it was in order to situate himself more firmly at the Academy of Sciences that he composed what passed for an autobiography and communicated it to the pertinent authorities.[6] His most valued patron was Ivan Shuvalov, while Leonhard Euler and Christian Wolff, by dint of both real and exaggerated association, were cherished patronage resources. It is, however, the characteristics that Lomonosov chose when fashioning his identity as a Russian scientist, along with

---

[5]   Biagioli, *Galileo, Courtier*. Biagioli's influence on Guiliano Pancaldi's exploration in biography, *Volta: Science and Culture in the Age of Enlightenment* (Princeton: Princeton University Press, 2003), is evident. Humphry Davy's very public efforts to fashion and re-fashion his identity, which often invited ridicule and accusations of superficiality from his many critics, are discussed in Jan Golinski, "Humphry Davy: The Experimental Self," *Eighteenth-Century Studies* 45, no. 1 (2011): 15-28.

[6]   Otto Sonntag's "The Motivations of the Scientist: The Self-Image of Albrecht von Haller," *ISIS* 65, no. 228 (September 1974): 336-51, explores the eighteenth-century Swiss-German natural philosopher Albrecht von Haller's evolving psychological "motivations" in striving to mold and promote his scientific status in Göttingen and Bern, which ranged from the religious to an incipient authorial self-interest, and nicely supplements Biagioli's more sociological approach. Haller's career was quite as encyclopedic and dependent on patronage as Lomonosov's, though the latter expressed little of the "ambivalence" towards earthly honors, or the "personal ambition and rivalry" needed to attain them, which Sonntag espies in Haller's writings.

the permutations that identity underwent after his death, that are mainly relevant to attempting to understand what he signified in Russian culture.

Seventeenth- and eighteenth-century natural philosophers— irrespective of country—revealed maddeningly little information about their inner lives for future biographers to utilize.[7] Lomonosov left a few direct references in his writings that later memoirists, litterateurs, historians, and scientists would use to great advantage in constructing an image of an extraordinarily diligent polymath, quite unique in time and place. Notably significant autobiographical reflections were conveyed in Lomonosov's letters to his well-placed Maecenas, Ivan Shuvalov (1727-97), a member of one of the more

---

[7]   Whether or not more contemporary scientists bequeath a fuller record of their non-working lives is debatable. A fascinating collection of articles exploring the genres of scientific biography and autobiography can be found in Shortland and Yeo, *Telling Lives in Science*. Studies of generally high quality in this same vein include A. Rupert Hall, *Isaac Newton: Eighteenth–Century Perspectives* (Oxford: Oxford University Press, 1999); Thomas L. Hankins, "In Defense of Biography: The Use of Biography in the History of Science," *History of Science* 17, no. 35 (March 1979): 1-16; Rosalyn D. Haynes, *From Faust to Strangelove: Representations of the Scientist in Western Literature* (Baltimore: Johns Hopkins University Press, 1994); Rebekah Higgitt, *Recreating Newton: Newtonian Biography and the Making of Nineteenth-Century History of Science* (London: Pickering & Chatto, 2007); Mary Jo Nye, "Scientific Biography: History of Science by Another Means?," *ISIS* 97, no. 2 (June 2006): 322-29; Dorinda Outram, "Scientific Biography and the Case of Georges Cuvier: With a Critical Bibliography," *History of Science* 14, no. 24 (June 1976): 101-37; Söderqvist, *History and Poetics*; and Yeo, "Images of Newton," 257-84. Encountering Newton's vast legacy of writings, Frank Manuel, striving to reconstruct Newton's personal life, seemed understandably frustrated when forced to admit that "His correspondence,... reveals him only by indirection; he kept no diaries, wrote no autobiography, left no intimate private notes about individuals among the millions of words of manuscript on all aspects of creation." Frank Manuel, *A Portrait of Isaac Newton* (New York: Da Capo 1968), 16. One of the exceptions of this tendency to autobiographical silence is Robert Boyle's account of his first sixteen years, *An Account of Philaretus during his Minority*. Authored when he was in his early twenties, it is reprinted in Michael Hunter, ed., *Robert Boyle by Himself and His Friends, with a Fragment of William Wotton's Lost 'Life of Boyle'* (London: William Pickering, 1994), 1-22.

powerful families of the day and a longtime favorite of the Empress Elizabeth.[8]

Lomonosov and Shuvalov are perhaps most famously joined in historical accounts by their efforts, largely led by Shuvalov, to found Moscow University in 1755.[9] Lomonosov was also persuaded by Shuvalov, or forced by the nature of his dependence on his patron, to abandon his science for long periods of time to engage in such work as assisting Voltaire in his writing of the *Histoire de l'empire de Russie sous Pierre le Grand* (which came out in two volumes in 1759

---

[8]   Lomonosov's use of patronage to advance his numerous professional objectives has not yet been subjected to a thorough study. E. V. Anisimov, Walter J. Gleason, Kirill Ospovat, and Viktor Zhivov have, however, begun the discussion with the following: Anisimov, "M. V. Lomonosov i I. I. Shuvalov," *Voprosy istorii estestvoznaniia i tekhniki*, no.1 (1987): 73-83; Gleason, *Moral Idealists, Bureaucracy, and Catherine the Great* (New Brunswick, NJ: Rutgers University Press, 1981), 24-33; Ospovat, "Lomonosov i 'pismo o pol'ze stekla': poeziia i nauka pri dvore Elizavety petrovny," *Novoe literaturnoe obozrenie*, no. 87 (2007): 148-83; idem, "Mikhail Lomonosov Writes to his Patron: Professional Ethos, Literary Rhetoric and Social Ambition," *Jahrbücher für Geschichte Osteuropas* 59, no. 2 (2011): 240-66; and Zhivov, "Pervye russkie literaturnye biografii kak sotsial'noe iavlenie: Trediakovskii, Lomonosov, Sumarokov," *Novoe literaturnoe obozrenie*, no. 25 (1997): 47-53. Ospovat's articles reason persuasively that Lomonosov's ceaseless determination to sculpt an acknowledged "social niche" in elite circles was, in fact, his primary occupation. For surveys of Ivan Shuvalov's career, see E. V. Anisimov, "I. I. Shuvalov — deiatel' rossiiskogo prosveshcheniia," *Voprosy istorii*, no. 7 (July 1985): 94-104; P. I. Bartenev, "I. I. Shuvalov," *Russkaia beseda* 1, part 6 (1857): 1-80; "Shuvalov, Ivan Ivanovich," in *Russkii biograficheskii slovar'*, vol. 23 (St. Petersburg, 1911; reprint, New York, 1962), 476-86. Shuvalov's nephew, Prince F. N. Golitsyn, composed an interesting eulogy to his late uncle following his death in 1797, "Zhizn' ober-kamergera Ivana Ivanovicha Shuvalova," which was eventually published in *Moskvitianin*, no. 6 (1853): 87-98.

[9]   Stepan Shevyrev, *Istoriia Imperatorskogo Moskovskogo universiteta, 1755-1855* (Moscow, 1855), 7-22. Shevyrev's remains the best study of the university's founding and early years. Lomonosov assertively offered advice, rarely heeded, on the establishment of Moscow University's gymnasium(s) to Shuvalov, as demonstrated in E. E. Rychalovskii, ed., *Istoriia Moskovskogo universiteta (vtoraia polovina XVIII – nachala XIX veka). Sbornik dokumentov. Vol. 1: 1754-1755* (Moscow, 2006); and I. P. Kulakova, *Universitetskoe prostranstvo i ego obitateli: Moskovskii universitet v istoriko-kul'turnoi srede XVIII veka* (Moscow, 2006).

and 1763),[10] as well as in writing two historical tracts of his own: *A Short Russian Chronicle with a Genealogy* (*Kratkii Rossiiskii letopisets s rodosloviem*) and *Ancient Russian History from the Beginning of the Russian Nation to the Death of the Great Prince Iaroslav I, or to 1054* (*Drevniaia Rossiiskaia istoriia ot nachala rossiiskogo naroda do konchiny velikogo kniazia Iaroslava Pervogo ili do 1054 goda*).[11] Lomonosov dedicated several works to Shuvalov; perhaps the best known of these is the *Letter on the Usefulness of Glass* (*Pis'mo o pol'ze stekla*, 1752).[12]

Lomonosov wrote more often and in greater detail to Shuvalov than he did to any other correspondent—between 1750 and 1764 he sent Shuvalov at least 34 letters. Through these letters, many of which were first published in the six-volume 1784–87 Academy of Science's edition of Lomonosov's collected works,[13] Lomonosov established the vague outlines of what would become constants in

---

10   See Carolyn H. Wilberger, "Voltaire's Russia: Window on the East," *Studies on Voltaire and the Eighteenth Century* 164, ed. Theodore Besterman et al. (1976): 23-133, for an examination of Voltaire's composition of *Histoire de l'empire de Russie sous Pierre le Grand*.

11   These works, which were first issued in 1760 and 1766 respectively, along with other works which might be deemed historical—such as Lomonosov's critical reaction to Voltaire's history—and an extensive commentary on their composition can be found in Lomonosov, *PSS*, vol. 6, 19-373, 541-95.

12   Ibid., vol. 8, 508-22, 1003-008. For its wide dispersion in the eighteenth century, see *Svodnyi katalog*, vol. 2 (Moscow, 1964), 163-66, 176; and Kuntsevich, *Bibliografiia izdanii sochinenii Lomonosova*. Lomonosov's letter in verse was part of a campaign by him to enlist Shuvalov's continuing aid in his efforts to build a factory for the production of colored glass.

13   *Polnoe sobranie sochinenii Mikhaila Vasil'evicha Lomonosova, s priobshcheniem zhizni sochinitelia i s pribavleniem mnogikh ego nigde eshche ne napechatannykh tvorenii*, vol. 1 (St. Petersburg, 1784), 319-45; idem, *PSS*, vol. 10, 468-587, 807-77, passim; and Martynov, *Lomonosov: perepiska*, passim. For a cogent, though dated, explication of the fate of Lomonosov's broader correspondence and the uses to which it has been put, see L. B. Modzalevskii's commentary to Lomonosov, *Sochineniia M. V. Lomonosova*, vol. 8 (Moscow-Leningrad, 1948), 5 40. With the exception of a communication of disputable attribution (see E. S. Kuliabko, "Neizvestnoe pis'mo I. I. Shuvalova k M. V. Lomonosovu," *XVIII vek* 7 [1966]: 99-105), only one letter from Ivan Shuvalov to Lomonosov has been found: Martynov, *Lomonosov: perepiska*, 209-10.

his historiography: tales of a mythic youth in the far north of Russia; his journeying for education to Moscow and then to Marburg, following a winding path to and through these cities in his search for the intellectual benefits he might receive; and his long years of heroic toil at the Academy of Sciences. Most interesting in these letters are those themes that would become biographical tropes in the elaborate mythology devoted to Lomonosov: his obstacle-strewn path to the sciences and the arduous, yet historically triumphant, nature of his labors once he arrived. Struggles engaged in are a presence throughout Lomonosov's representation(s) of his life.

There are two direct references of substance in Lomonosov's writings to Shuvalov pertaining to his childhood, one regarding his journey from Kholmogory to Moscow and the other in reference to his time spent at the Slavo-Greco-Latin Academy. Because of their authoritative status in later studies as Lomonosov's own ruminations, these references will be excerpted at some length. Lomonosov's letters are stylistically complex, even turgid, and personal details he conveyed were, as is to be expected, heavily bound up in questions of patronage and his own evolving self-identification. That contemporaries knew the content of both letters, that they were among those published in the first volume of Lomonosov's 1784-87 collected works, makes them particularly valuable.

In a letter of 1753 to Shuvalov, much of which was concerned with outlining some of his research on electricity and his experiments with a thunder machine (*gromovaia mashina*) performed together with fellow Academician Georg Richmann, Lomonosov began by profusely thanking Shuvalov, who, unlike the patrons of apparently unworthy fellow scientists, always asked for and received work of the highest quality from him.[14] For him, presumably as opposed to many others at the Academy of Sciences, the desire to learn,

---

[14]   Lomonosov, *PSS*, vol. 10, 480-482; and *Polnoe sobranie sochinenii Lomonosova*, vol. 1, 1784, 326-30. The letter is dated 31 May 1753. Whenever addressing Shuvalov or other patrons, Lomonosov's avowal of thanks was necessarily extravagant, and the language elaborately mannered, as the style of the period required.

the need for hard work, and the obligation to search for the truth were characteristics that he had exhibited since his youth, and he declared that:

> although my father was by nature a kind man, he was without learning while my stepmother was wicked and jealous, and at every opportunity she sought to anger him against me by saying I was lazy, satisfied only to waste away my time with books. Therefore, I found it necessary, again and again, to find a place to read and study in dark and desolate places, to suffer cold and hunger, until the time I was able to leave for the Spasskii school [the Zaikonospasskii Monastery, the home of the Slavo-Greco-Latin Academy].[15]

Lomonosov insisted that despite these deprivations, there was nothing to be ashamed of in his childhood. Quite the contrary, it would seem. Considering the hardships into which he was born, his present standing was even more astounding.

Lomonosov's miraculous rise from humble beginnings on the periphery of Russia; his early love of learning; his attraction to books, the titles of which later biographers would adduce with some creativity; and his journey to enlightenment, or at least what passed for enlightenment in eighteenth-century Russia—all staples of his biography—make their appearance for the first time in his letters. Lomonosov's passage from Kholmogory to Moscow and then the Slavo-Greco-Latin Academy has the aura of legend in both Russian and Soviet historiography. Enemies, even of a familial variety, are also present in Lomonosov's remembrances; omnipresent adversaries and obstacles overcome are a constant in the narratives.[16] From such thin autobiographical lore were myths constructed.

---

15   Lomonosov, *PSS*, vol. 10, 481-82.

16   Iurii Lotman and Boris Uspenskii compellingly argue that a manichean opposition was both present in and in fact necessary to the formation of Russian myths. The positive, almost godlike, qualities invested in the hero permitted no intermediary ground that might be shared with the antithetical anti-hero. See their "The Role of Dual Models."

Due to Lomonosov's strenuous efforts to garner Imperial backing for his proposed glass factory near St. Petersburg, which were finally rewarded by the court in early 1753,[17] his financial situation had become complicated, and the insufficiency of the support conferred by the state onto scientific endeavors, or rather their organizers, was at the forefront of his thinking even after he received the funding. He signaled his disquiet at the scant official largesse for Russian science in a May 1753 letter to Shuvalov,[18] in which he also responded to his benefactor's belief, or jest, apparently conveyed in an earlier note, that having been granted his request for a factory by the government he might now pursue his other scientific activities with less passion than he had previously demonstrated.

Lomonosov pointed out that if, despite his many past travails, his pursuit of knowledge had never been affected, then it certainly could not be so now, "for even when I lived in the utmost poverty, which for the sciences I willingly endured, I could not be deterred."

> When I was studying at the Spasskii school there were very strong influences from all sides to turn me away from learning, and these proved to be nearly irresistible. On the one hand, Father, not having any other children than myself, said that I, being his only child, had deserted him and all of the property and income (according to local conditions) which

---

[17]   Relevant documents pertaining to Lomonosov's quest, ultimately successful, to found and then maintain a factory for the manufacture of "colored glass," along with allied labors to create a mosaics factory, are located in Lomonosov, *PSS*, vol. 9, 73-181, 682-717. The great expenses involved in glass and mosaic production would eventually lead to Lomonosov's near financial ruination. See also V. K. Makarov, *Khudozhestvennoe nasledie M. V. Lomonosova: Mozaiki* (Moscow-Leningrad, 1950).

[18]   Lomonosov, *PSS*, vol. 10, 478-80; and *Polnoe sobranie sochinenii Lomonosova*, vol. 1, 1784, 324-26. Lomonosov's letter is dated 10 May 1753. Lomonosov likely portrayed the penury of his student days in Moscow with great accuracy. As for his being too old at twenty to commence with the study of Latin, however, this was not at all an unusual age to begin such studies at the Slavo-Greco-Latin Academy. Indeed, due to the school's chronic shortage of students, there was a wide variety in the ages and abilities of those admitted into the Academy. See S. K. Smirnov, *Istoriia Moskovskoi slaviano-greko-latinskoi akademii* (Moscow, 1855).

he had built up for me by his own sweat and blood. All of it, he said, would be seized by strangers after his death. On the other hand, in the Academy, I had to endure the most extreme poverty: I had only one *altyn* per day stipend, and could not spend more than half a *kopeck* for bread and half a *kopeck* for *kvas*; the rest was for paper, shoes, and other necessities. In this way I lived for five years and never gave up on learning. On the one hand, knowing my father's means, the people at home hoped to marry their daughters off to me, just as they had when I lived there. On the other hand, at the school many of the other pupils, who were young children after all, would point at me and yell, "Look at what a blockhead to start studying Latin at twenty."[19]

Lomonosov was also thankful for the opportunity to travel abroad to continue his studies, and held that support for the sciences, and the individual scientist, in other countries was munificent by comparison to the support available in Russia.

Referring to the comparatively comfortable professional lives enjoyed by Newton, Boyle, Hans Sloane and Wolff, he suggested that these scientists succeeded so spectacularly in part because they had been freed, in varying manners, from financial worry. These and other eminent and well-rewarded worthies (Leonhard Euler must be included on the list) were models to Lomonosov of people whose commitment to science was sustained by society. Lomonosov and his biographers from the eighteenth century until the present day have also made repeated analogies between his reputation and that of Benjamin Franklin. Lomonosov not only held up the achievements of these scientists for his patrons to examine, but he explicitly connected himself to their attainments. To the eighteenth-century scientific practitioner, especially one in such a socially insecure setting as Lomonosov's, known recognition by eminent worthies in the "Republic of Letters," as Enlightenment Europe may be termed,[20] provided the foundations for more securely establishing their honor.

---

19   Lomonosov, *PSS*, vol. 10, 479.

20   Gary Marker's "Standing in St. Petersburg Looking West, Or, Is Backward-

Although Lomonosov greatly exaggerated the level of state or regime maintenance available to his archetypal natural philosophers in Western Europe, which was arguably no greater than that to be found in Russia, the exaggeration was a useful rhetorical device. He was arguing for elevating the status of the natural philosopher in Russia; also implicit in his plea is the supposition that few could hope to match his own skills in surmounting the obstacles he had faced. Lomonosov's chronicle of the tribulations he faced as an inquisitive boy and then as a student in various locales fit perfectly the narrative pattern of travelers' accounts. The hardships he experienced on his journeys are necessarily magnified. The impediments would become even more severe in the interpretations of later memoirists.

In her work on the autobiographies of eighteenth-century French scientists, Dorinda Outram explores the dominant metaphor of "travel and becoming," of "life as a curious exploration of many paths."[21] This metaphor "allowed the linkage of the life and work to go on being made at another level not by cutting out the life but by seeing it as a web of movement, curiosity and introspection which came together in a scientific vocation." In considering the not atypical example of Georges Cuvier, Outram argues that for Cuvier, in remembering his early life, "the inner movement from childhood to adolescence was also a movement from one language to another, all encapsulated in an actual journey from his birthplace in provincial Montbeliard, to school in cosmopolitan Stuttgart."[22] Continuous conflict between Cuvier and the world that

---

ness All There Is?," *Republic of Letters: A Journal for the Study of Knowledge, Politics, and the Arts* 1, no. 1 (May 2009), is a spirited insistence on Russia's (and Lomonosov's) inclusion in the Republic of Letters.

[21]   Dorinda Outram, "Life-Paths: Autobiography, Science and the French Revolution," in Shortland and Yeo, *Telling Lives in Science*, 89-90.

[22]   Ibid., 89. See also Outram's more comprehensive work, *Georges Cuvier: Science, Vocation and Authority in Post-Revolutionary France* (Manchester: University of Manchester Press, 1984). Although focusing primarily on twentieth-century "science biographies," Thomas Söderqvist provides a perceptive study of the dialectic between the "production of knowledge" and the scientist's searching for a self-image in his "Existential Projects and

he entered—he described his passage as lined with all manners of obstacles—punctuated his autobiography. This journeying to enlightenment was especially difficult for prospective scientists who needed to radically realign their previous lives with subsequent self-representations. Lomonosov's ascent from peasant's son to the Academy of Sciences was acutely sharp, and the need for him to understand and explain this transition was commensurate with its acuity.[23]

The pilgrimage or journey has been a motif in diverse literatures for many centuries. During the Middle Ages, a traveler's or pilgrim's linear movement toward a "sanctified" destination was set in opposition to the morally suspect temptations of a secular curiosity, which continually threatened to divert the pilgrim from his journey. In his discussion of Richard de Bury's *Philobilon*, Chaucer's *Canterbury Tales* and Mandeville's *Travels*, Christian Zacher indicates that by the end of the Middle Ages pilgrimage had, however, "become little more than a mask concealing natural human yearnings to explore other lands—the journey itself, more than the sacred goal, became the objective of men's travels."[24]

Curiosity, the seeking of knowledge for its own sake, had fused with the journey; by the time of the Renaissance, they were no longer antithetical. "Pilgrims outlived the Middle Ages," Zacher

---

Existential Choice in Science: Science Biography as an Edifying Genre," in Shortland and Yeo, *Telling Lives in Science*, 45-84.

[23]   "Myth's plot as a text is very often based on the hero's crossing the border of a 'narrow' closed space and his passage into the external boundless world. However, it is precisely the notion of the availability of a small 'world of proper names' that lies at the basis of such plots' generative mechanisms. This sort of mythological plot begins with a passage into a world where the names of objects are unknown to man…. The very existence of an 'alien' open world in myth implies the presence of 'one's own' world, which is endowed with the feature of measurability and is filled up with objects bearing proper names." Lotman and Uspenskii, "Myth-Name-Culture," 237-38. The affinity between Lomonosov's self-fashioning in his letters to Shuvalov and mythological texts are distinct.

[24]   Christian K. Zacher, *Curiosity and Pilgrimage: The Literature of Discovery in Fourteenth-Century England* (Baltimore: Johns Hopkins University Press, 1976), 5.

stresses, "and curiosity remained a vice beyond the Renaissance—but wayfarers in different dress now clogged the roads, and if their business seemed curious it was curious largely in the modern sense of the word."[25] Such later works as John Bunyan's *Pilgrim's Progress*, which argued powerfully against a pilgrim's straying from his path, testified, however, to the residual strength of the conflict between curiosity and pilgrimage.[26]

This genre of pilgrimage and journeying accounts began to appear in Old Russian literature in the twelfth century, or roughly coinciding with the origins of literature in Medieval Rus', and is referred to as *khozhdeniia*.[27] Although "Western" and Russian models during the Middle Ages were in many ways profoundly distinct from one another, the tension expressed between secular and spiritual motivations was an increasingly shared one. By the eighteenth century, the divergences between Russian and West European forms, along with the internal tensions within the genre, were quickly disappearing.[28] Lomonosov's writings provide

---

[25]   Ibid., 16.

[26]   Outram, "Life-Paths," 88. John Bunyan's influence on eighteenth-century Russian literature has not been scrutinized, and it has yet to be demonstrated that Lomonosov ever came into contact with *Pilgrim's Progress*, even, for example, during his student days at Marburg, the time when he had his only sustained exposure to foreign literature. Despite its nineteenth-century emphasis, Dmitrii Blagoi's "Dzhon Benian, Pushkin i Lev Tolstoi," does touch on the eighteenth century. See D. D. Blagoi, *Ot Kantemira do nashikh dnei*, vol. 1 (Moscow, 1972), 334-40. See also Valentin Boss, *Milton and the Rise of Russian Satanism* (Toronto: University of Toronto Press, 1991), 3, 59. *Pilgrim's Progress* was translated from both French and German and published three times in Russia during the 1780s. The essayist and publisher Nikolai Novikov, author of a biography of Lomonosov, issued two of these editions. See *Svodnyi katalog*, vol. 1 (Moscow, 1962), 91; and V. P. Semennikov, *Knigoizdatel'skaia deiatel'nost' N. I. Novikova i tipograficheskoi kompanii* (Petrograd, 1921), 84-87.

[27]   Gail Diane Lenhoff Vroon, "The Making of the Medieval Russian Journey" (Ph.D. diss., University of Michigan, 1978), 1-17. The Byzantine roots of the *khozhdeniia* are carefully argued for in Lenhoff's dissertation (see especially pp. 22-49).

[28]   See Andreas Schönle, *Authenticity and Fiction in the Russian Literary Journey, 1790-1840* (Cambridge, MA: Harvard University Press, 2000), 1-6, for a

a telling example. While there is a decided similarity between his remembrances and the "conversion moments" of a pilgrimage in spiritual memoirs,[29] this is a largely a reflection of the continuing influence of earlier religious journeying patterns in literature. The objects around which Lomonosov structures his descriptions reveal a mainly secular sensibility.

In a letter of 4 January 1753 to Shuvalov, written at a time when he was engaged in historical studies at the apparent expense of his scientific labors, Lomonosov wrote of his many obligations:

> As for my other occupations in physics and chemistry, there is neither the need nor the possibility that I forsake them. Every man requires relaxation from his labors; for that purpose he leaves serious business and seeks to pass the time with guests or members of his household at cards, draughts, and other amusements, and some with tobacco smoke, which I had given up long ago, since I find nothing but boredom in them. And thus I hope that I shall be allowed several hours a day to relax

wider analysis of this trend in the Russian travelogue.

[29] Dorinda Outram outlines how "Moments of epiphany, of absorption in Nature, in scientific autobiography have the same role as conversion moments in spiritual autobiography: they resolve the antagonism of the self and the world." Outram, "Life-Paths," 93. For conceptions of pilgrimage, sacred spaces, and the mutable, inverted nature of the spiritual and profane values attached to them in eighteenth-century Russia, see Iu. M. Lotman and B. A. Uspenskii: "K semioticheskoi tipologii russkoi kul'tury XVIII ve-ka," in A. D. Koshelov, ed., *Iz istorii russkoi kul'tury*, vol. 4 (XVIII—nachalo XIX veka) (Moscow, 1996), 442-45; and idem, "The Role of Dual Models," 25. Lotman and Uspenskii trace the model of journeys, or "movements in space" toward "enlightenment," in the Russian imagination to Peter the Great's "Great Embassy" to the West in 1697-98. See also K. V. Sivkov, "Puteshestviia russkikh liudei za granitsu v XVIII veke" in *Puteshestviia russkikh liudei za granitsu v XVIII veke* (St. Petersburg, 1914), 5-9. Pilgrimages to the West were undertaken by quite a number of eighteenth-century Russians, among them, of course, Lomonosov. His broader journey within Russia, however, both literal and allegorical, from Kholmogory to the Academy of Sciences, was also plainly just such a "movement in space," and, perhaps, one of a far more exceptional kind. A. Iu. Andreev's valuable *Russkie studenty v nemetskikh universitetakh XVIII—pervoi poloviny XIX veka* (Moscow, 2005), surveys the experiences of Russian students in the most common destination during the eighteenth and nineteenth centuries, German-speaking lands.

from the labors which I have expended on the collection and composition of Russian history and on the beautification of the Russian tongue so that I may use these hours, rather than for billiards, for experiments in physics and chemistry, they serve not only as a replacement for amusement, but furnish exercise instead of medicine and can bring no less benefit and honor to my native land, than my first occupation.[30]

This passage became one of the most frequently reprinted extracts from all of Lomonosov's writings. Indeed, it would be difficult to find an account of him over the past two centuries that does not either quote it or allude to it. The tension between Lomonosov's work in chemistry and physics and the manifold other duties imposed on and sought by him is the ostensible subject of the letter. Many later examinations of Lomonosov explain the unfinished quality of his scientific labors as resulting less from his undisciplined work habits or gaps in his theoretical knowledge than from the onerous requirements of patronage, which prevented him from completing his researches in chemistry and physics.[31]

It would then seem to follow that untold discoveries were never made, or were postponed for later generations to make,

---

[30]   Lomonosov, *PSS*, vol. 10, 475; and *Polnoe sobranie sochinenii Lomonosova,* vol. 1, 1784, 322-24.

[31]   This theme is reiterated, with varying degrees of subtlety, in recent biographies of Lomonosov. See Evgenii Lebedev, *Lomonosov* (Moscow, 1990) (Lebedev's book came out in an enormous print run of 150,000 copies); G. E. Pavlova and A. S. Fedorov, *Mikhail Vasil'evich Lomonosov, 1711-1765* (Moscow, 1986); Rudolf Balandin, *Mikhail Lomonosov* (Moscow, 2011); Peter Hoffmann, *Michail Vasil'evic Lomonosov (1711-1765): Ein Enzyklopädist im Zeitalter der Aufklärung* (Frankfurt am Main: Peter Lang, 2011); O. D. Minaeva, *"Otechestva umnozhit' slavu...": biografiia M. V. Lomonosova* (Moscow, 2011); Shubinskii, *Mikhail Lomonosov;* and Serman, *Lomonosov.* Most of these authors also make judicious note in their works, however, of the tangible rewards that patronage brought to Lomonosov. Henry Leicester (*Lomonosov on the Corpuscular Theory,* 10) assigns partial blame for Lomonosov's peripatetic ways to his famous tendency to get bogged down in bitter disputes with other academicians. See also Vucinich, *Science in Russian Culture: A History to 1860,* 112-13. Conflict is, of course, part and parcel of the workings of patronage (as well as of institutionalized academic life).

because Lomonosov was forced to engage in non-scientific work. Such beliefs are, or course, far too speculative to be subject to serious scrutiny. What is unmistakable is that Lomonosov purchased Shuvalov's support for his chemical and physical science exertions by completing any commission, in any domain, that Shuvalov required of him. That Shuvalov deigned to subsidize and encourage his scientific activities indicates that an association with science brought some adornment to its sponsors. Patronage and Lomonosov's molding of his specific scientific self were inextricably bound together.

Conspicuous in Lomonosov's letter is the image of his selfless devotion to the sciences; a calling that served as a glorious respite from the toils that absorbed his daily life. His eighteenth-century biographers would greatly amplify this perception of his disinterest in any activities that might distract him from his intellectual pursuits. Eventually, the mantra in studies of Lomonosov as a scientist would be that his primary work—his real "first occupation"—was physics and chemistry and not the composition of history, the writing of odes and oratorical prose, developing linguistic investigations, or any of the sundry other ventures that competed for his attention. Although for a scientist to be active in a multiple array of fields was not viewed as unusual in Lomonosov's era,[32] it became untenable in the nineteenth century, and Lomonosov's brief comments concerning the division of his day would serve as a rationale to

---

[32]   Antoine Lavoisier typified the encyclopedic eighteenth-century savant whose career, seemingly split into myriad disparate professional roles, has often frustrated later efforts to organize it into a coherent, if artificial, whole. Jean-Pierre Poirier, *Lavoisier: Chemist, Biologist, Economist*, trans. Rebecca Balinski (Philadelphia: University of Pennsylvania Press, 1996), is an explicit, and somewhat successful, effort to address this issue. Marco Beretta, *Imaging a Career in Science: The Iconography of Antoine Laurent Lavoisier* (Canton, MA: Science History Publications, 2004), underlines the impediments, visual and textual, that prevent Lavoisier from being easily perceived as anything but a scientist. For an interesting deciphering of the many-sided career ot Joseph Banks (longtime president of the Royal Society), and how succeeding depictions of him encompassed, or failed to encompass, all of these dimensions, see Gascoigne, "The Scientist as Patron and Patriotic Symbol," 243-65.

later chroniclers for severing his activities into, principally, either science or literature. In the eighteenth century, however, his citing of his multiple roles should be viewed mainly as augmenting his idealized self-portrait.

Lomonosov's entreaties to Shuvalov soliciting greater respect for the scientist were also attempts to bring shape to a new and ill-defined social category in eighteenth-century Russia, that of the scientific practitioner.[33] Lomonosov's declaration that his work would bring honor and benefit to his "native land" was one that would have considerable appeal to later nationalist-minded historians of Russian science. With its strong assertion of his worth to Russia, however, it is better interpreted as an attempt to firmly prescribe his own position within the Academy of Sciences. He was the first indigenous Russian scientist to be made a full member of the Academy, and this nascent vocation was, initially through his own efforts, utterly conflated with his drive to elevate his self-

[33]   There are no satisfactory synthetic works dealing with the place of the scientist in pre-revolutionary Russian society. The situation of the chemist, however, is in part covered in Nathan Marc Brooks, "The Formation of a Community of Chemists in Russia: 1700-1870" (Ph.D. diss., Columbia University, 1989). Karl Hufbauer's influential prosopographical study *The Formation of the German Chemical Community (1725-1795)* (Berkeley: University of California Press, 1982) shapes both Brook's methodology and his conclusions. For an instructive essay that underscores the difficulties in establishing and maintaining a purely scientific vocation in eighteenth-century Russia, see R. W. Home, "Science as a Career in Eighteenth-Century Russia: the Case of F. U. T. Aepinus," *Slavonic and East European Review* 51, no. 122 (January 1973): 75-94. See also V. K. Novik, "Akademik Frants Epinus (1724-1802): kratkaia biograficheskaia khronika," *Voprosy istorii estestvoznaniia i tekhniki*, no. 4 (1999): 4-35; and J. Scott Carver, "A Reconsideration of Eighteenth-Century Russia's Contributions to European Science," *Canadian-American Slavic Studies* 14, no. 3 (Fall 1980): 389-98. Examinations of social structures in Imperial Russia that argue, with varying levels of success, for a quite amorphous, indeed "polymorphic," composition of society include Gregory L. Freeze, "The *Soslovie* (Estate) Paradigm and Russian Social History," *American Historical Review* 91, no. 1 (February 1986): 11-36; and Elise Kimerling Wirtschafter, *Social Identity in Imperial Russia* (DeKalb, IL, 1997), 3-99. Of course, the more porous the social boundaries, the easier it would have been for Lomonosov to reshape his status.

representation. This was a crucial element in the ascription of honor to his calling,[34] be it chemist, physicist, litterateur, or historian. Those aspiring to any of these new socioprofessional identities would not be priests in the temple as it were, but rather were aiming for systematized recognition of a wholly new kind.

That Lomonosov's status and that of the scientist in general were still quite fragile in Russia, however, is seen in two missives that he sent to Shuvalov. In 1754, he petitioned his patron to assist him in obtaining either a transfer out of the Academy of Sciences or a promotion to directing it himself. If granted the latter change, which he obviously would have preferred, he could put a stop to the "crafty undertakings" that were plainly damaging its operations.[35] If he left, though, he was convinced that posterity would regard him, and the Academy, as the victims, and predicted

> that all should say: "the stone that the builders rejected has become the chief cornerstone, this is the Lord's doing" … [and] in my departure from the Academy it will become perfectly clear of what it is losing, when it is deprived of such a man, who for so many years has embellished it, and has always fought against the persecutors of learning, despite the dangers to himself.[36]

---

34   Steven Shapin and Otto Sonntag consider, respectively, efforts by Robert Boyle and Albrecht von Haller to define not only their own honor and status, but that which might be accorded to the early modern natural philosopher in England and the Germanies. See Steven Shapin, *A Social History of Truth: Civility and Science in Seventeenth Century England* (Chicago: University of Chicago Press, 1994), 126-92; and Sonntag, "Motivations of the Scientist," 336-51.

35   Lomonosov, *PSS*, vol. 10, 518-19 (the letter's dating is 30 December 1754); and *Polnoe sobranie sochinenii Lomonosova*, vol. 1, 1784, 338-39.

36   Lomonosov, *PSS*, vol. 10, 519. Following Catherine II's ascension to the throne in 1762, and with Ivan Shuvalov—who was soon to leave Russia—no longer in a position to assist him in his undertakings, Lomonosov, possibly wishing to test his support at Catherine's court, applied to be discharged from the Academy of Sciences. In his application, reminiscent of the above letter, he reminds the empress of his valuable years of service to the Academy and of the great renown he has brought to it in wider scientific circles (ibid., 351). Lomonosov's request was, after much to and fro, dismissed; indeed, in 1764,

Lomonosov's supplication, as far as can be ascertained, was ignored. It is unlikely that he meant it to be seriously considered; his strategy was to persistently call attention to his plight and, whenever possible, to tie his fate securely to the perceived fortunes of the Academy of Sciences, and moreover to that of the sciences in Russia as a whole.[37]

The other letter was sent some years later when, incensed at a presumed slight, or "insult," by Baron Aleksandr Stroganov (later president of the Academy of Arts), Lomonosov wrote to Shuvalov that

> highly-placed people scorn me for my low origins, looking at me as if I were an eyesore, irrelevant of the fact that I won my honors not by blind chance, but, given to me by God, and for science, by my talent, diligence, and tolerance for suffering in extreme poverty.[38]

These two letters are remarkable illustrations of Lomonosov's laying claim to rank. He believed that his achievements were such that men of lesser accomplishment, even a noble like Stroganov, owed

---

to much fanfare, the Empress Catherine II even visited the scientist in his laboratory on the Moika. There, she viewed his mosaic art and "observed physics instruments that he had invented as well as several experiments in physics and chemistry." *Sanktpeterburgskie vedomosti*, no. 48 (1764). It is surprising that despite Catherine II's recognized disdain for Lomonosov, she honored him with a visit. As Joachim Klein intriguingly wondered in a private communication, was Catherine II's visit to Lomonosov not perhaps an early manifestation of a "cult of Lomonosov"? Klein alludes to the scale of the Soviet cult of Lomonosov in his *Russkaia literatura v XVIII veke* (Moscow, 2010), 99.

[37]   Lomonosov drew up several plans for the re-organization of the Academy of Sciences in the 1750s and early 1760s. Some of these proposals envisioned him at the helm of the Academy, in the position of vice-president. This post would not come into existence until 1800, and even then would exist only temporarily; its first holder would be Lomonosov's one-time foe, Stepan Rumovskii. Lomonosov also sought for the inclusion of the professors of the Academy in the "Table of Ranks." The more formal of Lomonosov's "reform" plans are located in Lomonosov, *PSS*, vol. 10, 11-167. Incessant complaints about mismanagement at the Academy, and proffered solutions, are also scattered throughout much of his correspondence.

[38]   Ibid., 539. Lomonosov wrote this letter on 17 April 1760.

him a level of deference. To enforce this required that the position of professor of chemistry at the Academy of Sciences, occupied by this humble fisherman's son, be fashioned into one held in some esteem. The style of the letters is familiar; they resonate with similarities to memoirs of journeying and discovery. Obstacles and enemies, as is Lomonosov's recollection of his younger days, are prominently displayed. Tacit in his writing is his faith in his eventual ascendancy over his rivals.

The letters to Shuvalov are not of consequence as guides to the minutiae of Lomonosov's scientific labors; rather, they are rare autobiographical sources portraying his ascent to scientific heights. One further letter is interesting in that it combines some detail on Lomonosov's actual science with an incident in his biography that added immeasurably to the legend that grew up around him.

Georg Richmann, a physicist and member of the St. Petersburg Academy of Sciences[39] who was Lomonosov's collaborator in experiments on electricity with a thunder machine, was killed on 26 July 1753 by a bolt of lightning. This event attracted enormous attention throughout Western Europe and America.[40] Lomonosov

---

[39]   G. K. Tsverava, *Georg Vil'gel'm Rikhman (1711-1753)* (Leningrad, 1977), 125-48, and A. A. Eliseev, *G. V. Rikhman* (Moscow, 1975), 94-109, outline Lomonosov's collaboration with Richmann on electricity in the months before Richmann's death and provide detailed reconstructions of the experiment that killed him. For Richmann's scientific papers—some of them previously unpublished—along with selected correspondence, see G. V. Rikhman, *Trudy po fizike*, eds. A. A. Eliseev et al (Moscow, 1956).

[40]   For an example of the reaction to Richmann's death, see Charles Rabiqueau's pamphlet *Lettre élèctrique sur la mort de M. Richmann* (Paris, 1753); and William Watson, "An Answer to Dr. Lining's query Relating to the Death of Professor Richman [sic]," *Philosophical Transactions of the Royal Society* 48 (1753): 765-72. Lomonosov's associations with Richmann are briefly noted in both accounts. Richmann's death received far more detailed attention outside of Russian than within the country. Benjamin Franklin printed an account of Richmann's death in *The Pennsylvania Gazette* (Leonard Labaree et al, eds., *The Papers of Benjamin Franklin*, vol. 5 [New Haven, CT: Yale University Press, 1962], 219-21). Franklin sanguinely concluded from Richmann's death that "The new Doctrine of Lightning is, however, confirm'd by this unhappy Accident; and many Lives may hereafter be sav'd by the Practice it teaches. M. Richmann being about to make Experiments on the Matter of Lightning,

conveyed a poignant description of Richmann's death, composed on the day of the accident, to Shuvalov.[41] His plea that this tragedy "should not be interpreted in a way that is injurious to the augmentation of the sciences," together with his professed determination to preserve Richmann's memory, was tailor-made for hagiography.

The circumstances of Richmann's death not only led to the contemporary celebration of his achievements but also inspired increased attention directed at the scientific labors of his colleagues, in particular, of course, those of Lomonosov at the Academy of Sciences. This last factor was absolutely vital in securing further support for both the individual scientists and for the "new" institutions and disciplines with which they were increasingly associated, for Lomonosov was wholly successful in joining his name and ostensible achievements in the study of electricity to those of the martyred, and then better-known, Richmann.

Lomonosov's work on electricity was the scientific research to which his early biographers most often referred. It was perhaps less theoretically weighty than most of his other writings, and had a definite potential for practical application. I would also argue that electrical experimentation provided a crucial early site for the theatrical mediation of knowledge (natural philosophical or naturalistic) from a specific group to an ever wider audience via demonstrations.[42] This audience was by far wider than what could be

---

had supported his Rod and Wires by Electrics *per se,* which cut off their Communication with the Earth; and himself standing too near where the Wire terminated, help'd with his Body to compleat that Communication. It is plain the Wire conducted the Lightning to him thro' the whole Length of the Gallery: And had his Apparatus been intended for the Security of his House, and the Wire (as in that Case it ought to be) continued without Interruption from the Roof to the Earth, it seems more than probable that the Lightning would have follow'd the Wire, and that neither the House nor any of the Family would have been hurt by that unfortunate Stroke."

[41]   Lomonosov, *PSS,* vol. 10, 484-85. Lomonosov's letter concerning Richmann was originally published in *Polnoe sobranie sochinenii Lomonosova,* vol. 1, 1784, 330-33.

[42]   For the fusion of theatricality and natural philosophy at the St. Petersburg

garnered by either scientific publications or public lectures (both of which the Academy of Sciences undertook, to yawning indifference, in St. Petersburg), and this "demonstration culture" would remain the dominant means of disseminating scientific knowledge in Russia throughout the eighteenth century. Arguably even more important to the academicians at the Academy of Sciences than the diffusion of knowledge from such episodes of "public science"[43] was that the corresponding blending of public spectacle, indeed showmanship, wide publicity, delighted patrons, and "science" wove together ever more tightly the web of patronage mechanisms on which their status and hence their livelihoods depended.[44]

Lomonosov's continuation of his and Richmann's experiments, dangerous though they had proven to be, fit perfectly into the heroic image expected of, and being written for, these early natural philosophers. That Richmann died not only during a moment of experimentation, in itself valuable in validating the apparently innovative methodology of the "new sciences," but during experiments with electricity, was especially valuable, for the study of electricity had become the emblematic "new science"[45] of

---

Academy of Sciences, see Simon Werrett, *Fireworks: Pyrotechnics Arts and Sciences in European History* (Chicago: University of Chicago Press, 2010), 103-31.

[43]   On how the "early modern" natural philosopher attempted to fabricate a receptive public, interesting introductions are Larry Stewart, *The Rise of Public Science: Rhetoric, Technology, and Natural Philosophy in Newtonian Britain, 1600-1750* (Cambridge: Cambridge University Press 1993); David N. Livingstone, *Putting Science in its Place: Geographies of Scientific Knowledge* (Chicago: University of Chicago Press, 2003); and Jan Golinski, *Science as Public Culture: Chemistry and Enlightenment in Britain, 1760-1820* (Cambridge: Cambridge University Press, 1992). Stewart and Golinski contribute chapters on the performance of science, on a variety of stages, to Bernadette Bensaude-Vincent and Charlotte Vincent, eds., *Science and Spectacle in the European Enlightenment* (Burlington, VT: Ashgate, 2008).

[44]   Perhaps the most influential work on the "social construction" of science is Steven Shapin and Simon Schaffer, *Leviathan and the Air Pump: Hobbes, Boyle, and the Experimental Life* (Princeton: Princeton University Press, 1985).

[45]   John Heilbron, *Electricity in the 17th and 18th Centuries: A Study of Early Modern Physics* (Berkeley: University of California Press, 1979), is the orthodox

the time. In Lomonosov's autobiography, and in the accounts of his contemporary biographers, the details of his scientific work were not yet as important as his path toward, and devotion to, science.

That the eighteenth century was an age of imitation is a commonplace, and Lomonosov's presentation of his past fits the pattern. As Iurii Lotman posits in his writings on the astonishing power of roles in Russian culture, "eighteenth-century man would choose a certain type of behavior for himself that simplified his real everyday life-style and elevated it to a certain idealized form."[46] Lotman concentrates on eighteenth-century Russian literary typologies, providing a catalogue of theatricalized models in which Lomonosov plays a central part. He was persuaded that the manifold transformations that eighteenth-century Russia underwent, launched by Peter the Great,[47] or later in his name,

---

study of the topic. Less portentous, but more provocative, are Patricia Fara, *An Entertainment for Angels: Electricity in the Enlightenment* (Cambridge: Icon Books 2002); and James Delbourgo's inspired *A Most Amazing Scene of Wonders: Electricity and Enlightenment in Early America* (Cambridge, MA: Harvard University Press, 2006).

[46]   Lotman, "The Poetics of Everyday Behavior," 241-242. See also Lotman and Uspenskii, "The Role of Dual Models," 18-28; Marcus C. Levitt, introduction to Marcus C. Levitt, ed., *Dictionary of Literary Biography*, vol. 150, *Early Modern Russian Writers: Late Seventeenth and Eighteenth Centuries* (Detroit, 1995), XIII-XVI.

[47]   The literature on Peter's reign is immense; however, a concise introduction to several of the Petrine "revolutions" is offered in James Cracraft, ed. *Peter the Great Transforms Russia*, 3rd ed. (Lexington, MA: D.C. Heath and Co., 1991). Lindsey Hughes's *Russia in the Age of Peter the Great* is a thorough study of the nature and depth of Peter's reforms/revolutions. The tensions in Petrine historiography, implicit in defining specific changes as either as reforms or revolutions, or neither, are also systematically examined (see in particular her concluding chapter, "The Legacy," pp. 462-70). For an investigation of the "cultural revolution" directed by Peter the Great, see James Cracraft's three-volume consideration of the topic: *The Petrine Revolution in Russian Architecture* (Chicago: University of Chicago Press, 1988); *The Petrine Revolution in Russian Imagery* (Chicago: University of Chicago Press, 1997); and *The Petrine Revolution in Russian Culture* (Cambridge, MA: Harvard University Press, 2004). See also Richard S. Wortman, *Scenarios of Power: Myth and Ceremony in Russian Monarchy*, vol. 1 (Princeton: Princeton University Press, 1995); V. M. Zhivov, *Iazyk i kul'tura v Rossii XVIII veka*

were often experienced as cataclysms that utterly annihilated traditional patterns of life and ushered in a radically new age. This made Russians especially prone to a remaking of the self based on a limited number of cultural markers.

The introduction of science into eighteenth-century Russia, a development initiated by Peter the Great, inarguably brought about a revolutionary alteration in the lives of a number of Russians, and the images of scientists, with that of Lomonosov the primary signifier, were among those heroic myths which gave assistance to those attempting to negotiate a new age. The details of scientists' work were projected onto their life stories beginning in the mid-nineteenth century, a time when an emergent scientific community and the interested public sought for the science in the lives of these representative subjects.

Prior to that time, it was the scientist's biography, as composed of certain stock heroic features, which was of principal import. Lomonosov's dissertations, partiuclarly those in physics/mechanics and chemistry, were assessed by contemporaries, many of them quite illustrious figures in scientific circles of the time.[48]

(Moscow, 1996); and idem, "Kul'turnye reformy v sisteme preobrazovanii Petra I," in *Razyskaniia v oblasti istorii i predystorii russkoi kul'tury* (Moscow, 2002), 381-435. Evgenii Anisimov's *Dyba i knut: politicheskii sysk i russkoe obshchestvo v XVIII veke* (Moscow, 1999) appears to sum up much of that historian's efforts to highlight the highly coercive nature of Peter's reign. This in itself, in his estimation, was a momentous transformation, and one that provided the institutional and perhaps even intellectual bases for state repression in Russian and later Soviet society.

[48] Lomonosov's meteoric rise to "scientific greatness" is typically portrayed as not only symbolic of, but indeed confirmation of, the Petrine revolution in matters scientific. For Peter the Great's interest in and encouragement of the sciences, see Valentin Boss, *Newton and Russia: The Early Influence, 1698-1796* (Cambridge, MA: Harvard University Press, 1972), 9-96; Robert Collis, *The Petrine Instauration: Religion, Esotericism and Science at the Court of Peter the Great, 1689-1725* (Turku, Finland: University of Turku Press, 2007); Iu. Kh. Kopelevich, *Osnovanie Peterburgskoi Akademii nauk* (Leningrad, 1977); N. I. Nevskaia et al., eds., *Letopis' Rossiiskoi Akademii nauk*, vol. 1, *1724-1802* (St. Petersburg, 2000), 15-30; K. V. Ostrovitianov, *Istoriia Akademii nauk SSSR*, vol. 1 (Moscow-Leningrad, 1958), 30-56; P. P. Pekarskii, *Nauka i literatura v Rossii pri Petre Velikom*, vol. 1, *Vvedenie*

These assessments, however, were often either sharply critical of Lomonosov's suppositions (and hence of no value to Lomonosov as patronage tools) or extended decidedly vague, though flattering, commentary on Lomonosov's promise, and thus could be adapted in a creative manner by Lomonosov.

Evaluations of Lomonosov by Christian Wolff (1679-1754) and Leonhard Euler (1707-83) were especially valuable "instruments of credit"[49] in Lomonosov's strategies for advancement as a natural philosopher or scientist, and had a critical bearing on subsequent scholarship. Wolff and Euler both enjoyed long and fruitful ties with Russia and with the St. Petersburg Academy of Sciences. Wolff served as a successor of sorts to Leibniz in providing advice to Peter the Great in planning for the Academy, and was also helpful in recruiting the first class of professors.[50] Despite invitations proffe-

---

v istoriiu prosveshcheniia v Rossii XVIII stoletiia (St. Petersburg, 1862); idem, Istoriia Imperatorskoi Akademii nauk v Peterburgie, 2 vols (St. Petersburg, 1870-73); Ludmilla Schulze, "The Russification of the St. Petersburg Academy of Sciences and Arts in the Eighteenth Century," British Journal for the History of Science 18 (1985): 305-35; K. Sharf, "Osnovanie Berlinskoi Akademii nauk i ikh otnosheniia v XVIII v. v evropeiskoi perspektive," in Nemtsy v Rossii: tri veka nauchnogo sotrudnichestva, ed. G. I. Smagina (St. Petersburg, 2003), 7-38; I. V. Tunkina, "Lavrentii Lavrent'evich Bliumentrost," in V. S. Solovev, ed. Vo glave pervenstvuiushchego uchenogo sosloviia Rossii: ocherki zhizni i deiatel'nosti prezidentov Imperatorskoi Sankt-Peterburgskoi Akademii nauk. 1725-1917gg. (St. Petersburg, 2000), 13-28; Alexander Vucinich, Science in Russia Culture: A History to 1860 (Stanford: Stanford University Press, 1963); and Simon Werrett, "An Odd Sort of Exhibition: The St. Petersburg Academy of Sciences in Enlightened Europe" (Ph.D. diss., Cambridge University, 2000). Pekarskii's works remain the most comprehensive survey of Russian science under Peter the Great. Michael D. Gordin's "The Importance of Being Earnest: The Early St. Petersburg Academy of Sciences," ISIS 91, no. 1 (March 2000): 1-31, is a tentative effort to unravel what may have been Peter's wider societal intentions when planning the Academy of Sciences. Gordin, in a nod to Norbert Elias, sees the Academy as central to a "civilizing process" launched by the Emperor.

49   On "credit," patronage, and career advancement, see Mario Biagioli, Galileo's Instruments of Credit: Telescopes, Images, and Secrecy (Chicago: University of Chicago Press, 2006), an extension of his earlier Galileo, Courtier (Chicago: University of Chicago Press, 1993).

50   Kopelevich, Osnovanie Peterburgskoi Akademii nauk, 65-79. Kopelevich

red to him to join the nascent Academy, Wolff never visited Russia. Euler, by contrast, spent many productive years at the Academy (1727-41, 1766-83), and after a quarter of a century at the Berlin Academy would return to St. Petersburg with great honors by invitation of Catherine II.[51] The mere association of Lomonosov

---

contests the importance of Wolff's help in setting up the Academy. She does not, however, provide enough evidence to appreciably change the reader's perception of Wolff's role. Lavrentii Blumetrost's singular contributions to founding the Academy of Sciences are also elided in Kopelevich's account. For a corrective, see Eduard Winter, "L. Blumentrost d.J. und die Anfänge der Peterburger Akademie der Wissenschaften. Nach Aufzeichnungen von K. F. Svenske," in *Jahrbuch für Geschichte der UdSSR und der volkdemokratischer Länder Europas*, vol. 8 (1964): 247-69. Liselotte Richter, *Leibniz und sein Russlandbild* (Berlin: Deutsche Akademie der Wissenschaften zu Berlin, 1946), gives Leibniz excessive intellectual credit for the Academy project. On Leibniz's early contacts with Peter the Great, see pp. 42-52 especially. While Richter's was for a long time the most detailed work on Leibniz's Russian connections; her work can now be supplemented, though not replaced, by N. P. Kopaneva and S. B. Koreneva, *G. V. Leibnits i Rossiia* (St. Petersburg, 1997). D. Meiers's chapter on Leibniz and the Petersburg Academy of Sciences is both concise and authoritative. On Leibniz's correspondence with Peter the Great (and more notably with various advisors and courtiers to the Emperor in Russia, or otherwise knowledgeable about Russia across Europe), see Vladimir Ger'e (W. Guerrier), *Leibniz in seinen Beziehungen zu Russland und Peter dem Grossen* (St. Petersburg, 1873).

51   Euler's legacy is awarded profuse attention in numerous admiring collections: A. M. Deborin, ed., *Leonard Eiler, 1707-1783: sbornik statei i materialov k 150-letiiu so dnia smerti* (Moscow-Leningrad, 1935); M. P. Lavrent'ev, A. P. Iushkevich, and A. T. Grigor'ian, eds., *Leonard Eiler: sbornik statei v chest' 250-letiia so dnia rozhdeniia, predstavlennykh Akademii nauk SSSR* (Moscow, 1958); N. N. Bogoliubov, G. K. Mikhailov, and A. P. Iushkevich, eds. *Razvitie idei Leonarda Eilera i sovremennaia nauka: sbornik statei* (Moscow, 1988); Robert E. Bradley and C. Edward Sandifer, eds., *Leonhard Euler: Life, Work and Legacy* (Amsterdam: Elsevier, 2007); and V. N. Vasil'ev, ed., *Leonard Eiler: K 300-letiiu so dnia rozhdeniia. Sbornik statei* (St. Petersburg, 2008). See also Boss, *Newton and Russia*; Ronald S. Calinger, "The Introduction of the Newtonian Natural Philosophy into Russia and Prussia (1725-1772)" (Ph.D. diss., University of Chicago, 1971); idem, "Leonhard Euler: The First St. Petersburg Years (1727-1741)," *Historia Mathematica* 23 (1996): 121-66; Iu. Kh. Kopelevich, "Leonard Eiler—deistvitel'nyi i pochetnyi chlen Peterburgskoi Akademii nauk," in *Akademechiskaia nauka v Sankt-Peterburge v XVIII-XX vekakh: istoricheskie ocherki*, ed. E. A. Tropp and G. I. Smagina (St. Petersburg, 2003), 55-72; Rudol'f Mumentaler, *Shveitsarskie uchenye v*

with such famed natural philosophers greatly enhanced his standing at the time and for posterity.

Over time Christian Wolff became a far more difficult figure
to fit into representations of Lomonosov's scientific genius than was
Euler.[52] Wolff's prestige in eighteenth-century thought, primarily

---

Sankt-Peterburgskoi akademii nauk (St. Petersburg, 2009), passim; and Eduard
Winter, ed., *Die deutsch-russische Begegnung und Leonhard Euler: Beiträge zu
den beziehungen zwischen der deutschen und der russischen Wissenschaft und
Kultur im 18. Jahrhundert* (Berlin: Akademie-Verlag, 1958). Petr Pekarskii's
short biography of Euler, part of his history of the early Academy of
Sciences, reflects, as does his work on Lomonosov, a sure command of
primary sources. See Pekarskii, *Istoriia Imperatorskoi Akademii nauk*, vol. 1,
247-308. No single biography of Euler exists that succeeds in conveying
the astonishing range of his activities. Emil A. Fellmann, *Leonhard Euler*
(Hamburg: Rowohlt Taschenbuch, 1995) is probably the best of the life and
letters variety. Its brevity (125 pages), however, undermines its interpretive
scope.

[52] Among the better studies of Lomonosov's tutelage under Christian Wolff
are M. I. Sukhomlinov, "Lomonosov—student Marburgskago universiteta,"
Russkii vestnik 31, no. 1 (January 1861): 127-65; and especially A. A. Morozov,
M. V. Lomonosov: Put' k zrelosti, 1711-1741 (Moscow-Leningrad, 1962), 221-
304. Neither of them, however, deals sufficiently well with the vexing
issue of the extent of Wolff's *long-term* impact on Lomonosov's scientific
work. V. A. Zhuchkov, ed., *Khristian Vol'f i filosofiia v Rossii* (St. Petersburg,
2001) attempts to rehabilitate Wolff's reputation; it is, unfortunately,
marred by a pedestrian (and historiographically thin) reading of Wolff's
influence over Lomonosov. Leopold Auburger, *Russland und Europa: Die
Beziehungen M. V. Lomonosovs zu Deutschland* (Heidelberg: Groos, 1985),
23-28, offers a brief survey of Lomonosov's associations with personages
and matters "German," though Wolff's influence over Lomonosov's
natural philosophy is alleged rather than substantiated. Wolff's travails in
attempting to administratively oversee Lomonosov and two fellow students
(G.–U. Raiser and D. I. Vinogradov) who accompanied him from St. Peters-
burg, along with Wolff's views of Lomonosov's abilities, are concisely set
down in Peter Scheibert, "Lomonosov, Christian Wolff und die Universität
Marburg," in *Academia Marburgensis: Beiträge zur Geschichte der Philipps-
Universität Marburg*, ed. Walter Heinemeyer, Thomas Klein, and Hellmut
Seier (1977), 231-40. For Wolff's early relationship with the Petersburg
Academy, particularly significant is Günter Mühlpfordt, "Deutsch-russische
Wissenschaftsbeziehungen in der Zeit der Aufklärung. Christian Wolff
und die Gründung der Petersburger Akademie der Wissenschaften," in
450 Jahre Martin-Luther-Universität Halle-Wittenberg, vol. 2, ed. Leo Stern
(1952), 169-97.

as the leading disseminator of Leibniz's natural philosophy, was immense, though it was already in some decline by the middle of that century (in Western Europe, not in Russia). The mystical and obfuscatory connotations that marked his name, and Voltaire's attacks on him, were especially devastating in this regard (his most withering taunts of the Wolffians are evident in *Candide*),[53] made later linkages with him problematic. Wolff's strong opposition to Newton, whose work he quite simply failed to grasp, greatly reduced his luster in the eyes of later generations of Russian scientists and historians of science. Although this aspect of Lomonosov's scientific makeup is in much dispute, it seems clear that Lomonosov never rose above the low level of mathematics that he took from his years (1736-39) as a student of Wolff's at Marburg. Justin Duising, a chemist at Marburg, was also, briefly, Lomonosov's teacher in mathematics. Without a deeper study of mathematics than Lomonosov seems to have undertaken at Marburg, any understanding of, let alone collaboration with, the most exciting scientific advances of the time was impossible.

---

[53]   Pangloss, the didactic teacher of "metaphysico-theologo-cosmolonigology" in *Candide*, who taught that "there is no effect without a cause," represents a stinging caricature of Christian Wolff. See Voltaire, *Candide*, trans. Donald M. Frame (reprint, New York: New American Library, 1981), 16. Voltaire's indignation at Wolff's "scientific" speculations was often expressed in his correspondence. In a letter of 10 August 1741 to Maupertuis, printed in Theodore Besterman et al., eds., *Les Oeuvres Complètes D Voltaire*, vol. 92 (Geneva: Institut et Musée Voltaire, 1970), 95, Voltaire famously ridiculed Wolff's theorizing about the size of the inhabitants of Jupiter: "Il y avoit longtemps que j'avoit vu avec une stupeur de monade, quelle taille ce bavard germanique assigne aux habitans de Jupiter. Il en jugeoit par la grandeur de nos yeux, et par l'éloignement de la terre au soleil. Mais il n'a pas l'honneur d'être l'inventeur de cette sottise, car un Volfius met en trente volumes les inventions des autres, et n'a pas le temps d'inventer." See also W. H. Barber, *Leibniz in France: From Arnauld to Voltaire, 1670-1760* (Oxford: Clarendon Press, 1955), 178-97, 231-35; and Boss, *Newton and Russia*, 169. For Wolff's negligible impact as an expositor of Leibniz on natural philosophic discussions in France, consult J. B. Shank, *The Newton Wars and the Beginning of the French Enlightenment* (Chicago: University of Chicago Press, 2008), 427-29, 437-47. Voltaire's zeal to discredit Leibniz—whom he painted as the errant foe of Newton's perfect universe—explains his acidic, and injudicious, dismissal of Leibniz's self-appointed successor, Wolff.

After his return to Russia, Lomonosov maintained a great respect for Wolff, and in 1746 he translated and published the first part of L. F. Tümmig's *Institutiones philosophiae Wolfianae*.[54] Lomonosov's translation, *Wolffian Experimental Physics* (*Volfianskaia eksperimental'naia fizika*), reissued with a supplement by him in 1760, was one of his most frequently published pieces.[55] It has long been treated, though, less as an example of Wolff's continuing sway over Lomonosov than as a singular contribution to early Russian scientific terminology.[56] Lomonosov often referred with great opprobrium to those whose writings he utilized; and Wolff's name is cited supportively throughout Lomonosov's multitudinous scientific papers. The rehashed Leibnizian notions of "contradiction" and "sufficient reason"[57] that characterized Wolff's approach to natural philosophy were embraced by Lomonosov in his scientific disquisitions and never abandoned. Their essentially deductive

---

[54]   Lomonosov, *PSS*, vol. 1, 421-530, 577-92.

[55]   Lomonosov's supplement to the 1760 edition can be found in ibid., vol. 3, 434-39. For the relatively large print runs of the *Volfianskaia eksperimental'naia fizika*, see *Svodnyi katalog*, vol. 1, 186-87.

[56]   A very well-argued paper within this framework is V. V. Zamkova, "Fizicheskaia terminologiia v 'Volfianskoi eksperimental'noi fizike' M. V. Lomonosova," in *Materialy i issledovaniia po leksike russkogo iazyka XVIII veka*, ed. Iu. S. Sorokin (Moscow-Leningrad, 1965), 87-115. See also E. A. Tropp, "Fizika i khimiia M. V. Lomonosova," in *Lomonosov: sbornik statei*, vol. 10, 12-19.

[57]   Wolff's methodology was in large part based on the following principles: "I. Philosophy is the science of all possible things, together with the manner and reason of their possibility.... II. By Science I understand, that habit of the understanding whereby, in a manner not to be refuted, we establish our assertions on irrefragable grounds or principle.... III. I call possible, whatever thing can be, or whatever implies no contradiction, whether actually existing or not.... IV. As of nothing we can form no conception, so neither can we of the actual existence of any thing, without a sufficient ground or reason; ... V. A philosopher ought, therefore, not only to know the possibility of a thing, but also to assign the reason of that existence." This is from Wolff's *Vernünfftige Gedanken von den Kraften des menschlichen Verstandes und ihrem richtigen Gebrauche in Erkäntniss der Wahrheit* (1713), as cited in Leicester, *Lomonosov on the Corpuscular Theory*, 12. Wolff required that occurrences in nature be assigned a cause, and not "merely" described.

search for ultimate, or first, doctrines in the study of natural phenomena made it impossible for Wolff and Lomonosov to accept Newton's theories, particularly his principle of action at a distance in the workings of gravitation, as anything but a reintroduction of "occult qualities" into science.[58]

Lomonosov's treatises in physics and chemistry manifest a thoroughly mechanical articulation of the makeup of the solid state that had little in common with Wolff's metaphysics. His methodological assumptions, however, remained strongly Wolffian; thus leaving him ill-equipped to grapple with the "new philosophy." As Valentin Boss concludes, it was an implacable hostility to Newtonianism that was Wolff's "worst legacy" to his Russian student, and one from which Lomonosov was never able to free himself.[59]

For later scholars the key to associating Lomonosov with Wolff was to either ignore or reject any direct scientific connection between Wolff's natural philosophy and Lomonosov's later scientific work. Instead, the emphasis has been on highlighting Wolff's somewhat fragmentary evaluations of Lomonosov's general progress and potential. These very vague reports scattered throughout various communications Wolff sent to Baron Korff (the president of the

---

[58] Lomonosov rejected the idea that gravity could be innate within bodies; he considered this a return to discredited occult notions. In an interesting letter to Euler (written in 1748), he articulated his opposition to Newton's theory of gravity and proposed, in limited circumstances, the existence of a "gravitational fluid" that acted on corpuscles and drove them to "the center of the earth." Lomonosov's corpuscular viewpoint, from which he very rarely veered, did not admit the possibility of material bodies operating on each other without an intervening medium. Lomonosov, *Sochineniia*, vol. 8, 1948, 72-91 (2), 18-22; and idem, *PSS*, vol. 10, 439-57, 801. He later (1758) offered this letter as a paper, essentially unchanged and entitled *On the Quantitative Relationship of Matter and Weight* (*Ob otnoshenii kolichestva materii i vesa*), to the Academy of Sciences. Ibid., vol. 3, 349-71, 556-58.

[59] Lomonosov's unqualified failure to appreciate the revolutionary nature of Newton's philosophic and scientific conceptions "was to have unfortunate consequences for his scientific work as a whole." Boss, *Newton and Russia*, 164. Boss's monograph deals at considerable length with Lomonosov's views on Newtonian science.

Academy of Sciences) in St. Petersburg[60] were concerned mainly with his day-to-day life in Marburg, and did not contain any information on Lomonosov's first scientific treatises.[61] Wolff's comments that the three students (Lomonosov, G.-U. Raiser and D. I. Vinogradov) dispatched to Marburg were with some acumen studying mathematics and languages, were attending his courses on mechanics and natural history, and would soon study experimental physics with him,[62] provided just the right generalities that could later be much elaborated.

Wolff placed more hope in the successful progress of Lomonosov than of either Raiser or Vinogradov, both of whom seem to have had more troubles with their courses of study. As he noted to Korff in 1738: "Lomonosov is evidently the brightest of the group; with the diligence he possesses he could learn a great deal and he displays an eagerness and willingness to learn."[63] This

---

[60]   Wolff wrote several letters to Academy President Korff detailing the activities of Lomonosov, Raiser, and Vinogradov. These letters were first published in A. Kunik, ed., *Sbornik materialov dlia istorii Imperatorskoi Akademii nauk v XVIII veke,* part 2 (St. Petersburg, 1865), 253-342, passim.

[61]   Lomonosov completed two scientific treatises under Wolff's supervision: *Work in Physics on the Transformation of a Solid Body into a Liquid, Depending on the Motion of a Pre-Existing Fluid (Rabota po fizike o prevrashchenii tverdogo tela v zhidkoe, v zavisimosti ot dvizheniia predsushchestvuiushchei zhidkosti,* 1738); and *A Physical Dissertation on the Differences Among Mixed Bodies, Ascribed to the Cohesion of Their Corpuscles (Fizicheskaia dissertatsiia o razlichii smeshannykh tel, sostoiashchem v stseplenii korpuskul,* 1739). See Lomonosov, *PSS,* vol. 1, 7-63, 539-45. Both works were sent to the Academy of Sciences for review. His second, far more substantial, essay dealt with the structure and nature of matter; it was an early sample of his corpuscular work. Wolff's metaphysics, or more specifically, his version of Leibniz's monadology, did not infuse Lomonosov's treatise; rather it anticipated the rigorously mechanistic viewpoint that he would embrace throughout his working life. There is no evidence that either paper was ever presented formally at the Academy, and no contemporaneous critiques of them have come down to us. Boris Menshutkin translated the papers from the Latin and published them in 1936.

[62]   Kunik, *Sbornik materialov,* part 2, 258-59; Pekarskii, *Istoriia Akademii nauk,* vol. 2 (St. Petersburg, 1873), 291.

[63]   Kunik, *Sbornik materialov,* part 2, 271-72; Pekarskii, *Istoriia Akademii nauk,* vol. 2, 291-92.

meshed perfectly with the heroic tales of Lomonosov's youthful perspicacity which his eighteenth-century biographers further amplified.

Even Wolff's ambivalent assessments—he made many complaints to the authorities at the Academy in St. Petersburg regarding Lomonosov's, Raiser's, and Vinogradov's decidedly dissolute ways—had the potential to be recast in a favorable light. Soon after the students departed for Freiberg to continue their studies (July 1739), Wolff wrote to Korff of his relief that they were gone and no longer his concern. He had, much to his consternation, been forced to assume their debts on occasion. The cause for many of the students' difficulties, particularly their chronic indebtedness to local merchants, was that "they were excessively wild and had a weakness for the opposite sex."[64] Their boisterous ways had apparently caused no end of upset to the professor, and, it would seem, to the townspeople. Wolff might have been aghast at the students' personal conduct, but he also added, albeit almost as a postscript, that Lomonosov "especially enjoyed successes in his studies of the sciences." The rakish aspects of Lomonosov's life would later be seen more positively as a human complement to his scientific biography. Unrestrained behavior, which famously defined Lomonosov's public and private demeanor, was often constructed as a necessary escape from his laborious studies and afterwards from his exacting work in St. Petersburg.

Attempting to advance his election to an honorary membership of the Bologna Academy in 1764, Lomonosov wrote a letter to Count Mikhail Vorontsov (the one-time Russian Chancellor, then in Italy), who occasionally rendered significant support to him, in which he systematically outlined some of his scientific labors and contributions to Russian learning. He included a lengthy attachment with selected commentaries and recommendations of his work, "Testimonials Concerning the Sciences of Councillor Lomonosov"

---

64   Kunik, *Sbornik materialov*, part. 2, 305; Pekarskii, *Istoriia Akademii nauk*, vol. 2, 294-95.

("Svidetel'stva o naukakh sovetnika Lomonosova").[65] Translated by Lomonosov himself, and leading the list, was an evaluation of his progress by Wolff. It apparently was given to him at around the time he was leaving Marburg (July 1739), and is starkly different, at least in tone, from the report that he sent to Korff the same month. Wolff reported that:

> The very able young man, Mikhail Lomonosov, has diligently attended my lectures in mathematics, philosophy, and physics since his arrival in Marburg and has zealously attempted to acquire a clear knowledge of the fundamentals. I have no doubt that if he should continue with his studies with the same degree of diligence he may upon his return to his native land be of great benefit to his people, which I desire with all my heart.[66]

---

[65] P. P. Pekarskii, *Dopolnitel'nyia izvestiia dlia biografii Lomonosova*, (St. Petersburg, 1865), 93-8; Lomonosov, *Sochineniia*, vol. 8, 1948, 270-88, (2) 272-82; and idem, *PSS*, vol. 10, 396-404, 569-80, 787-90, 871-74 (sharp editorial variations distinguish these three collections). There were fifteen reviews, or fragments of reviews, in Lomonosov's package. See PFA RAN, f. 20, op. 3, no. 134, ll. 30-31, ll. 49-52ob., ll. 60-63; and op. 1, no. 5, ll. 153-58, for various copies made by Lomonosov, and by contemporaneous scribes, of the originals. For a skeletal outline of the document's composition and its dispatch to Vorontsov in Italy, see V. L. Chenakal et al., eds., *Letopis' zhizni i tvorchestva M. V. Lomonosova* (Moscow-Leningrad, 1961), 400-03. Lomonosov's overall correspondence to Vorontsov is available in Lomonosov, *PSS*, vol. 10, 477-583, passim; and Martynov, *Lomonosov: perepiska*.

[66] Pekarskii, *Dopolnitel'nyia izvestiia*, 93-94; Lomonosov, *PSS*, vol. 10, 571, 872. The original assessment, in Wolff's handwriting, has evidently not survived. One of the copies held by Lomonosov, which he dated 10 October 1760, is located in PFA RAN, f. 20, op. 3, no. 71, ll. 1-2. Accompanying Wolff's letter in the archive is a general recommendation of Lomonosov's studiousness and potential by his mathematics teacher at Marburg, Justin Duising (which was also apparently obtained by Lomonosov in July 1739). Dated 10 October 1760, it was not included among the evaluations sent to Vorontsov. Duising's reputation was perhaps insufficiently stellar for Lomonosov to employ it in higher-level patronage activities. The original appraisal by Duising is located in the Marburg University archives (Sukhomlinov, "Lomonosov — student Marburgskago universiteta," 158; and Andreev, *Russkie studenty v nemetskikh universitetakh*, 152).

This letter served as a precious reference for later scholars. The "diligence" which Lomonosov displayed in his study of the sciences, not clouded by precise details, was wonderful fodder for later memoirists to work with and would remain a key source for investigating what science he took with him from Marburg. The value implicit, both for Lomonosov and for two centuries of later studies, in Wolff's stated "desire" that Lomonosov might be of benefit to his native land is so obvious as to need no explication.

Lomonosov no doubt thought that this document would buttress his candidacy for the Bologna Academy. In this instance it was not necessary, for shortly before the arrival of his letter at the end of April 1764, he was elected to the Academy. However, Lomonosov's translation and use of Wolff's endorsement as late as 1764 testifies not only to his desire to be associated with Wolff's still potent reputation,[67] but also reflects remarkably well the self-perception he was strongly attempting to claim at the time. The emphasis on his success in the sciences while at a relatively young age complemented the scientific papers he cited with "personal" stories attesting to his abilities. While there might be, due to its dubious history, some question as to whether Wolff actually wrote

---

[67]   Some years earlier, in 1754, Lomonosov wrote to Leonhard Euler in Berlin that though his present theorizing would undermine the "mysticism" still existing in natural philosophy, a result he only welcomed, he was loath to publish his "proofs" for fear that they might bring offense to a certain "elderly gentleman" (Christian Wolff), to whom he owed a great deal. Leibnizian/Wolffian monads were, after all, one of the "mystical" currents in science that he believed his work exposed as fallacious. See Lomonosov, *PSS*, vol. 10, 503, 827-28. Incidentally, though in his student days Euler had been an admirer of Wolff, he rather soon thereafter became deeply opposed to Wolffian metaphysics. It was not only, or merely, the scientific foundations of monadology that were suspect to Euler; the apparent deism—or even "atheism"—that Wolff's (and Leibniz's) monadic conceptions seemed to furnish intellectual support to inspired his disquiet. Ronald Calinger discusses Euler's disillusionment with Wolffian doctrines in "The Introduction of the Newtonian Natural Philosophy into Russia and Prussia," 147-49, 167-91, 250-66 and "Leonhard Euler: The First Petersburg Years," 153-54. See also K. Grai, "Leonard Eiler i Berlinskaia Akademiia nauk," in Bogoliubov, *Razvitie idei Leonarda Eilera*, 85-89.

this report,[68] it was long accepted as genuine and put to use by both Lomonosov and his biographers, which is more pertinent to this monograph than its actual provenance.

In the only known letter to Lomonosov attributed to Wolff, the German philosopher expressed his "great pleasure" in reading his papers in "*Kommentarii*," the St. Petersburg Academy's scientific journal (*Novi Commentarii Academiae scientiarum imperialis Petropolitanae*, successor to *Commentarii Academiae scientiarum imperialis Petropolitanae*), pointing out the "great honor you have brought to your people," and wishing that he "might be an example for others to come."[69] That Lomonosov brought honor to his people, a hint of what might emerge from Russia, would become a cliché in later studies; this sort of thing is central, of course, to the presentation of any fabled scientist. In the search for the earliest possible origins of this idea, Wolff's brief remarks have been invaluable. Wolff the discredited foe of the Newtonian system was turned into an esteemed eighteenth-century personage who placed considerable store in Lomonosov's scientific potential. That Wolff's name could only add to Lomonosov's image was taken for granted in the eighteenth-century accounts of his life. In the nineteenth

---

[68]   Since Wolff had died in 1754, he would have been quite unable to challenge its veracity.

[69]   Wolff did not specify any particular scientific paper in this communication. The first two volumes of *Novi Commentarii*, issued in 1750 and 1751 respectively, included several dissertations by Lomonosov, among them what was at the time perhaps his best-known purely theoretical treatise, written in 1744, *Meditations on the Cause of Heat and Cold* (*Meditationes de caloris et frigoris causa/Razmyshleniia o prichine teploty i kholoda*). See Lomonosov, PSS, vol. 2, 7-55, 647-52. Lomonosov included Wolff's letter in the series of fifteen evaluations that he enclosed in his 1764 missive to Vorontsov in Italy. Again, the original of Wolff's letter has not been found; we have only Lomonosov's partial translation (in which the letter is dated 26 July 1753). See Pekarskii, *Dopolnitel'nyia izvestiia*, 94; Lomonosov, *Sochineniia*, vol. 8, 131-32, (2) 69-70; and idem, *PSS*, vol. 10, 571-72, 872. That neither this letter nor Wolff's previously referred-to review of Lomonosov's student years, both very useful to his "self-styling," appears to have been preserved might raise questions about their descent; at this point, no conclusions can be offered with certainty.

century, however, when scientific memoirs were structured more around experiments and discoveries and less around youthful genius and "life-paths,"[70] Wolff's connections with Lomonosov became something that had to be more artfully or subtly conveyed.

Despite his opposition to selected aspects of Newton's work, Leonhard Euler's scientific standing was never seriously challenged, and no negative connotations became attached to his name. His mathematical reputation rested on solid ground, and the publication in 1736 of his *Mechanica* brought enormous credit to him and to the St. Petersburg Academy of which he was a member. In large part, Euler laid the foundations for a long and honorable tradition in mathematics in Russia.[71] As a natural philosopher he had no equal at the Academy, and that such an illustrious personage was so closely associated with the early years of Russian science has made him sacrosanct in the history of that science. As a foreigner, however, Euler could never be the subject of the kind of indigenous mythmaking that characterized the evolution of Lomonosov's scientific reputation.

Euler left St. Petersburg on 8 June 1741 and only returned in 1766, the year after Lomonosov's death. So unless Lomonosov met Euler during his brief sojourn at the Academy in 1736 before his departure for Marburg, of which there is no proof, Lomonosov and Euler were never personally acquainted.[72] They did, however, enjoy

---

[70]   Outram, "Life-Paths," 98.

[71]   A. P. Iushkevich's work on early Russian mathematics remains the most exhaustive. For outlines of Euler's formidable contributions to what is perceived as a Russian mathematical tradition, see his "Eiler i russkaia matematika v XVIII v.," *Trudy Instituta istorii estestvoznaniia* 3 (1949): 45-116; and idem, *Istoriia matematiki v Rossii do 1917 goda* (Moscow, 1968), 103-215.

[72]   One of the more thoroughly researched studies exploring Lomonosov's relationship with Euler is V. L. Chenakal's "Eiler i Lomonosov," in Lavrent'ev, Iushkevich, and Grigor'ian, *Leonard Eiler: sbornik statei*, 423-63. Chenakal rather effectively maintains that Euler had an almost wholly favorable estimation of Lomonosov's scientific abilities. The author exaggerates, however, in claiming that a true scientific collaboration existed between the two of them. See also G. E. Pavlova, "Lomonosov v kharakteristikakh i vospominaniiakh sovremennikov," *Voprosy istorii estestvoznaniia i tekhniki*, no. 3 (1986): 62-4.

a limited correspondence, punctuated by long silences, over some fifteen years.[73] Lomonosov's letters largely concerned the substance of his scientific labors. He often described his activities and advanced new ideas he was working on to Euler, and awaited his judgment. His later quite famed note to Euler of 5 July 1748, in which he is said to have first presented his law on the conservation of matter in chemical transformations[74]—and therefore to have anticipated Antoine Lavoisier's discovery—is perhaps the most outstanding

---

[73]   Lomonosov wrote six known letters to Euler between 1748 and 1765. For the texts of these letters, see Lomonosov, *PSS*, vol. 10, 436-598, 799-866, passim. Lomonosov was the recipient of at least four letters from Euler between 1748 and 1754. They have been published in Lomonosov, *Sochineniia*, vol. 8, 1948, 69-185, (2) 15-124, passim. See also V. L Chenakal, "Novye materialy o perepiske Lomonosova s Leonardom Eilerom," in *Lomonosov: sbornik statei*, vol. 3 (Moscow-Leningrad, 1951), 249-59. Chenakal's article concentrates on the events leading up to and including the initiation of their correspondence. He also reproduces the texts of their first letters. Even given the graphomania that marked the work of many of the natural philosophers of his day, Euler was an unusually active correspondent. A partial record of his voluminous surviving letters is detailed in T. N. Klado et al., eds., *Leonard Eiler: perepiska, annotirovannyi ukazatel'* (Leningrad, 1967). Euler's wide patronage reach is testified to in this collection, as is how little, as compared with his writings to other luminaries of the day, and as opposed to the thrust of Russian and Soviet historiography, Lomonosov impacted on his scientific life. *Die Berliner und die Petersburger Akademie der Wissenschaften im Briefwechsel Leonhard Eulers*, A. P. Iushkevich and Eduard Winter, eds., with the collaboration of Peter Hoffmann (Berlin, 1959-1976), embraces nearly all of Euler's correspondence with the members of the St. Petersburg Academy. Euler's correspondence with Lomonosov is found in vol. 3. If its goals are successful, the Euler Society will in time place Euler's vast writings online (see *The Euler Archive*: http://www.eulerarchive.org/).

[74]   Although this letter was not published in its entirety until 1948, when it was accompanied by a Russian translation of the Latin original, it was commented on, or quoted, in works published as early as 1865. For the full text of the letter, see Lomonosov, *Sochineniia*, vol. 8, 1948, 72-91 (2), 18-22; and idem, *PSS*, vol. 10, 439-57, 801. For a highly skeptical review of claims that Lomonosov's research anticipated or influenced Lavoisier's findings, see Philip Pomper, "Lomonosov and the Discovery of the Law of the Preservation of Matter in Chemical Transformations," *AMBIX* 10, no. 3 (October 1962): 119-27. See also Henry M. Leicester, "Boyle, Lomonosov, and the Corpuscular Theory of Matter," *ISIS* 58, no. 192 (Summer 1967): 240-44; and idem, *Lomonosov on the Corpuscular Theory*, 44-47.

example in their correspondence. The contents of these letters were not of great interest to the early memoirists.[75] In its earliest manifestations, the mythology of Lomonosov as pioneer naturalist more easily assimilated Wolff's attestations of Lomonosov's potential to rise to the heights of learning than the messier minutiae of day-to-day scientific work exhibited in the Euler correspondence. Then, in the nineteenth century, when scientific memoirs focused far more on the presumed essence of their subject's deeds in science, the now long-discredited Wolff's associations with Lomonosov were, as already indicated, gradually finessed into near irrelevance, while the specifics of Euler's views became ever more fundamental to idealized portraits of Lomonosov.

This is not to suggest that Euler's appraisals of Lomonosov were earlier known only to the two parties involved, for Lomonosov determinedly employed Euler's assessments to secure and advance his status. Euler's opinions of his scientific worth were indispensable to him in his academic and personal struggles within the Academy; even though he was living abroad, Euler's was a potent voice in Russian science circles. Seven of the fifteen aforementioned commentaries on his scientific activities and

---

[75]   Lomonosov's communications with Euler, along with Euler's assessments of him to others, unlike many of Lomonosov's letters to Shuvalov, only began to be published—quite absent any substantive commentary—in the 1840s. While it is true that by the middle of the eighteenth century the correspondence of several scientists of Newton's time were already published (specifically that of Robert Boyle and that between Leibniz and John Bernoulli), Rupert Hall concludes that though they "threw fresh light on Newton and his times," it "would be long before biographers discovered the usefulness of such material." Hall, *Isaac Newton*, 6. The nature of scientific biographies was such that the more technical aspects within the letters could only with some difficulty be assimilated into the life of the subject. An exception was William Wotton's efforts to include some of Boyle's correspondence in his planned biography of the scientist. His never completed biography, which he worked on for more than a decade commencing in 1696, was, with his reliance on Boyle's unpublished papers in his analysis, an apparent attempt to move beyond the issuance of a "panegyric" to something approaching a study of his subject's intellectual life, and is meticulously investigated in Hunter, *Robert Boyle by Himself and His Friends*, XXXVI-LIV.

potential that Lomonosov sent to Vorontsov in 1764 were by Euler.[76] With two exceptions, they, like the review fragments ascribed to Wolff, were translated or copied by Lomonosov, and no longer exist in their original, presumably more complete, form. There is a certain formulaic quality to them, as is to be expected in such largely excerpted testimonial materials. One of Euler's evaluations, however, most marked him as a supporter of Lomonosov in the historiography, and warrants repeating.

Examining two of Lomonosov's papers, *A Dissertation on the Action of Chemical Solvents in General* (*Dissertatsiia o deistvii khimicheskikh rastvoritelei voobshche*, 1743) and *Physical Meditations on the Cause of Heat and Cold* (*Fizicheskie razmyshleniia o prichine teploty i kholoda*, 1744)[77] for possible publication in *Commentarii* (they would eventually be published in *Novi Commentarii* in 1750), Euler wrote on 10 November 1747 to Academy President Kirill Razumovskii that:

> All of these dissertations are not only good, but superior, because he writes about physical and chemical properties which are necessary, but which, until now, were unknown, and which the most intelligent people were unable to explain. He was able to achieve this with such sound arguments, that I am completely convinced of the precision of his proofs. At this time I must in all fairness to Mr. Lomonosov conclude that he is in possession of a fortunate capacity for delineating phenomena in physics and chemistry. It can only be wished that all the other academies [members of the Academy] could show the same resourcefulness as demonstrated by Mr. Lomonosov.[78]

[76] Pekarskii, *Dopolnitel'nyia izvestiia*, 94-8; Lomonosov, *Sochineniia*, vol. 8, 1948, 282-86; and idem, *PSS*, vol. 10, 572-78, 872-72.

[77] When *Fizicheskie razmyshleniia o prichine teploty i kholoda* (*De causis caloris et frigoris meditationes physicae*) was published, with some emendations, in *Novi Commentarii*, it went under the title *Razmyshleniia o prichine teploty i kholoda* (*Meditationes de caloris et frigoris causa/Meditations on the Cause of Heat and Cold*). For both the original and revised versions, see Lomonosov, *PSS*, vol. 2, 7-55, 63-103, 647-53.

[78] A. F. Vel'tman, "Portfel' sluzhebnoi deiatel'nosti M. V. Lomonosova," in

The last point especially Lomonosov must have found to be quite useful. Euler soon received a note from Lomonosov thanking him for his review, and attempting to establish an ongoing correspondence concerning his researches with him.[79] It is easy to see why Lomonosov found Euler's report appealing enough to preserve and later to send on to Vorontsov. Far less concerned with assessing the particulars of Lomonosov's papers than were many of the letters and commentaries between them, which were in any case favorable, Euler's statements in support of Lomonosov's work, implicitly casting him as a representative natural philosopher, indeed one to be emulated, fit well with the self-image that Lomonosov was tirelessly propagating.

Johann Schumacher, the director of the Academy of Science's Chancellery (although Schumacher had long held only the innocuous sounding title of librarian at the Academy, he had in fact overseen its operations since the late 1720s under a succession of often disinterested presidents),[80] in 1753 solicited Euler's opinion of Lomonosov's paper on electricity, *Discourse about Air Phenomena, Caused by Electricity* (*Slovo o iavleniiakh vozdushnykh, ot elektricheskoi sily proiskhodiashchikh*).[81] Lomonosov's findings had been challenged

---

Ocherki Rossii, book 2 (Moscow, 1840), 6-7. See also Pekarskii, *Dopolnitel'nyia izvestiia*, 94-95; Lomonosov, *Sochineniia*, vol. 8, 1948, 282; idem; *PSS*, vol. 10, 572-73. 872-73. Only a copy of the original—in French—most probably made by Lomonosov, or at any rate under his supervision, is extant, and it can be found in PFA RAN, f. 136, op. 2. no. 43, l. 1. This might yet again inspire doubts as to the editorial liberties that he might or might not have taken in his transcription. The more important dimension of Lomonosov's use of this extract, though, is that it unequivocally represents both his belief in his own expansive abilities and his efforts to make certain that others should be equally aware of his intellectual gifts.

[79]   Lomonosov, *PSS*, vol. 10, 436-38, 799-800 (letter is dated 16 February 1748).

[80]   For starkly opposing views of Schumacher, see Pekarskii's influential negative depiction in *Istoriia akademii nauk*, vol. 1, 15-65; and Simon Werrett's revisionist interpretation: "The Schumacher Affair: Reconfiguring Academic Expertise across Dynasties in Eighteenth-Century Russia," *Osiris*, vol. 25, no. 1 (2010): 104-26.

[81]   Lomonosov, *PSS*, vol. 3, 15-99, 512-22.

at the Academy assembly during which he first presented them, with professor of astronomy A. N. Grishov delivering the main rebuttal.[82] While questioning several aspects of the dissertation, Grishov found it especially curious that some of Lomonosov's theories—that electricity was correlated with the northern lights, and that ascending and descending air currents lead to or rather constituted electricity in the atmosphere—had, at least in part, been proposed earlier by Benjamin Franklin. Though he allowed that Lomonosov might not have known about Franklin's findings, he also implied that he should have.

Ill-disposed to accept any criticism,[83] Lomonosov soon became entangled in acrimonious disputes with Grishov and other academicians (especially N. I. Popov and I. A. Braun). Schumacher called on Euler's authority in order to settle the dispute. Adjudicating

---

[82]   Grishov's critique of Lomonosov's exposition on electricity can be found in N. I. Idel'son, "Teoriia Lomonosova o stroenii komet. Novye dannye k 'Slovu o iavleniiakh vozdushnykh, ot eliktricheskoi sily proiskhodia-shchikh' (26 noiabria 1753 g.)," in *Lomonosov: sbornik statei*, vol. 1, 81-82. Contemporary reports and correspondence connected with Lomonosov's electrical paper and its public presentation at the Academy of Sciences are found in Pavlova, *Lomonosov v vospominaniiakh*, 114-25.

[83]   Despite his acquaintance with Franklin's electrical experiments (Lomonosov obliquely cited Franklin's *Experiments and Observations on Electricity, Made at Philadelphia in America*, 1751), he disavowed any notion that he was indebted to his work, writing, "In my theory about the cause of electric power in the air I owe nothing to him," and professing that "I only saw Franklin's writings when I had already prepared my speech." After listing what he perceived to be gaps in Franklin's research, which Lomonosov argued resulted largely from Franklin's observations having been made in a far different climate than St. Petersburg's—Philadelphia's—he concluded, "there are many phenomena related to thunder [and related atmospheric changes], which in Franklin's work there are no traces of." See Lomonosov's 1753 paper, *Iz'iasneniia, nadlezhashchie k slovu o elektricheskikh vozdushnykh iavleniiakh*), in Lomonosov, *PSS*, vol. 3, 103 (101-133 for entire treatise). This paper was a supplement to his *Slovo o iavleniiakh vozdushnykh, ot elektricheskoi sily proiskhodiashchikh*, ibid., 15-99, 512-22. Also published in 1753, it was reprinted in the 1784-87 edition of Lomonosov's collected works consulted by Aleksandr Radishchev. For the several eighteenth-century incarnations of these two dissertations, see *Svodnyi katalog*, vol. 2, 163-66, 175-76.

the controversy, Euler barely touched on Grishov's report, though he dismissed it; rather he focused on the overall significance of Lomonosov's treatise, and on his role in Russian science:

> The composition of Mr. Lomonosov about this subject, I read with great pleasure. The explanations given by him in regard to the sudden onset of cold and its beginnings in the upper levels of the air in the atmosphere, I consider absolutely well founded. Not long ago I drew similar conclusions from the study of equilibrium in the atmosphere. His other suppositions are as intelligent as they are probable and display the author's successful talent for disseminating a true understanding of the natural sciences, other examples of which by the way are also manifested in his earlier works.[84]

The endorsement of Euler was vitally important in buttressing Lomonosov's position at the Academy. By seeming to tie himself so directly to Lomonosov's science (whose hypotheses, at least in the area of electricity, he did fundamentally agree with; after all, he had come to similar conclusions himself), Euler provided fortification for Lomonosov in intra-Academy disputes.

This is not to say that because of his support Lomonosov's theoretical findings triumphed, even in Russia, over alternative explanations. Indeed, in the study of electricity, Franz Aepinus's contemporaneous work in St. Petersburg quickly superseded Lomonosov's. Euler was so highly respected, however, that Lomonosov was able in part to disregard many of his critics in intra-Academy squabbles over position and status—though his temper seemed to be such that he did not. In light of Lomonosov's apparently angry reactions to reproofs regarding his electrical dissertation, which were causing much indignation at the Academy, Schumacher importuned Euler to reevaluate his stance toward

---

[84]   PFA RAN, f. 1, op. 3, no. 44, ll. 80-81 (Euler's letter to Schumacher is dated 29 December 1753). See also P. S. Biliarskii, *Materialy dlia biografii Lomonosova* (St. Petersburg, 1865), 248-49; Pekarskii, *Istoriia Akademii nauk*, vol. 2, 526-27.

Lomonosov, a call that was gently rebuffed.[85] He generally found little to disapprove of in Lomonosov's theories, but at the same time, he continued to receive financial support from the Academy, whose head was Schumacher. Euler plainly hoped to avoid being drawn

---

[85]   Biliarskii, *Materialy dlia biografii Lomonosova*, 251-52; Pekarskii, *Istoriia Akademii nauk*, vol. 2, 528; Chenakal, "Eiler and Lomonosov," 438-40. (Schumacher's letter was sent in early January 1754; Euler's answer came about a month later.) The battles that Lomonosov waged over the course of his academic career with the leadership of the Academy, particularly with Johann Schumacher and his son-in-law Johann Taubert, were bitter and historically became enveloped in the idea, with little evidence to corroborate it, that Lomonosov was fighting to advance the interests of Russian science against the intrigues of foreigners—read Germans—at the Academy. His struggles against entrenched, often non-Russian, elements within the administration of the Academy are a central theme of Radovskii's *Lomonosov i Akademiia nauk*. For observations about Lomonosov contained in the correspondence between Gerhard Friedrich Müller and Leonhard Euler, see E. Vinter and A. P. Iushkevich, "O perepiske Leonarda Eilera i G. F. Millera," in Lavrent'ev, *Leonard Eiler: sbornik statei*, 471-83. Müller, a historian and editor, was a longtime member of the Academy of Sciences. He knew both Euler and Lomonosov well, though his interactions with the latter became stormy. Müller is often named as one of the foreign academicians who did so much to hinder native Russian scholarship. J. L. Black's *G. F. Müller and the Imperial Russian Academy* (Kingston and Montreal, 1986), attempts to correct this view. The tercentenary of Müller's birth saw the publication of a major biography by Peter Hoffmann: *Gerhard Friedrich Müller (1705-83): Historiker, Geograph, Archivar im Dienste Russlands* (Frankfurt am Main: Peter Lang 2005). Hoffmann's work, along with *G. F. Miller i russkaia kul'tura* (St. Petersburg, 2007), edited by Dittmar Dahlmann and Galina Smagina (Peter Hoffmann is among the more than 40 contributors), is unrivaled in documenting Müller's contributions to eighteenth-century Russian culture. Müller's relationship with Lomonosov is treated at length in Hoffmann and Black's monographs; both assign much of the blame for the souring of Müller's and Lomonosov's once cordial, even friendly, associations on Lomonosov. Dahlmann and Smagina's collection was issued as part of a series emerging from a yearly seminar on German-Russian scientific and cultural contacts: *Nemtsy v Rossii: russko-nemetskie nauchnye i kul'turnye sviazi/ Die Deutschen in Russland: Russisch-deutsche wissenschaftliche und kulturelle Beziehungen Die deutschen in russland: russisch-deutsche wissenschaftliche und kulturelle Beziehungen*. This conference and seminar series is a noteworthy intellectual successor, of sorts, to the work begun by Eduard Winter and his "school" in the 1950s.

into the quarrels between Lomonosov and a burgeoning cast of other scholars.

Euler's epistolary relationship with Lomonosov soured in 1754-55, and was apparently never fully restored.[86] Lomonosov's last, extremely dyspeptic, letter to Euler (written at the end of February 1765) exists only in what looks to be an unfinished state, and it appears not to have been dispatched to Berlin. Lomonosov is scathing of those he perceived as foes at the Academy— Euler himself was addressed quite disrespectfully, while Stepan Rumovskii (Lomonosov's former student in Petersburg, and later a mathematician and professor of astronomy at the Academy of Sciences)[87] was referred to as the "lapdog" of Johann Taubert, Schumacher's successor as de facto head of the Academy, and a man whose abilities Lomonosov bitterly deprecated.[88]

---

[86]   In an attempt to answer critics of his work in the West, Lomonosov, without Euler's permission, published part of a supportive letter Euler had sent to him (dated 31 December 1754) in Le Caméleon Littéraire (St. Petersburg, no. 20, 18 May 1755, 453-56). See also Lomonosov, Sochineniia, vol. 8, 1948, 181-85, (2) 121-23. Euler's angry response, expressed to Müller, can be found in ibid., 124. Euler and Lomonosov were also on opposite sides of a dispute over the selection of a new professor of mathematics, experimental physics, and mechanics at the Academy. Euler promoted the candidacy of his student S. K. Kotel'nikov, while Lomonosov backed J. K. Spangenberg of Marburg. This somewhat puts paid to the notion of Lomonosov's encouragement of native scientists. Earlier on, Lomonosov dispatched a letter (7 May 1754) to Müller that was rather disparaging of Euler and Kotel'nikov. So that his relationship with Euler would not be "disturbed," Lomonosov cautioned Müller to make sure his comments did not come to his attention. See Lomonosov, PSS, vol. 10, 506-08.

[87]   V. V. Bobynin, "Rumovskii, Stepan Iakovlevich," Russkii biograficheskii slovar', (St. Petersburg, 1918; reprint, New York, 1962), 441-50; and G. E. Pavlova, Stepan Iakovlevich Rumovskii, 1734-1812 (Moscow, 1979). Long the director of both the Academy of Sciences' astronomical observatory (1763 to 1803) and its geographical department (1766-1803), Rumovskii had an astoundingly varied career in the sciences. Though he wrote original papers in mathematics, astronomy, and geography, he is probably known best for translating into Russian Euler's acclaimed Letters to a German Princess (1st ed. 1768-74).

[88]   The incomplete draft of the letter, in German, is in PFA RAN, f. 20, op. 1, no. 2, ll. 336-37. Lomonosov's note was published by Vel'tman in "Portfel'

Rumovskii had studied and worked with Euler in Berlin, and was firmly enrolled as his devotee. Whatever his initial thoughts might have been on Lomonosov's scientific adeptness, which cannot be determined, by the late 1750s he had become quite contemptuous of his forays into a variety of areas.[89] Perhaps it was for purely patronage and/or personal reasons that Rumovskii sided with Euler in his evident dispute with Lomonosov, or perhaps, as was entirely possible, he had concluded that Lomonosov's work was unsound. In any case, Rumovskii and the mathematician Semen Kotel'nikov were Euler's protégés, and came to represent his legacy, not Lomonosov's, in Russian science. They also supplanted Lomonosov as Euler's closest contacts among Russian scientists at the Academy. Despite this, Lomonosov understood the value of an association with Euler and did not tire in exploiting it, hence his use as late as 1764 of Euler's now rather aged evaluations, even though their frayed relationship was hardly restored.

Lomonosov's self-representation, supplied in his autobiographical epistles to Shuvalov, were structured around obscure origins followed by pathways to temples of learning, a journey made difficult every step of the way by various obstacles and foes, and, finally, the blessing received on arrival, scientific eminence. If full acceptance were not granted to him in the company of the learned, it would be by history. His success in linking his

---

sluzhebnoi deiatel'nosti Lomonosova," 69-72. It was eventually rendered into Russian and printed in *Zapiski Imperatorskoi Akademii nauk* 5, book 1 (1864): 105-06. Along with vituperative commentary on Rumovskii and Taubert, Lomonosov referred to Müller and the late Schumacher (who had died in 1761) with acutely vitriolic remarks.

[89]   Rumovskii wrote two letters to Euler, in 1756 and 1757, reacting negatively to some of Lomonosov's recent ideas. Lomonosov's new telescope (which was meant to be used at night, or during otherwise less than ideal conditions), his discussion of the proportionality of material and weight, or rather his inability to discuss it, and his faltering efforts to oppose Newton's concepts on gravitation, came in for considerable derision. See Pekarskii, *Dopolnitel'nyia izvestiia*, 74-79; and idem, *Istoriia Akademii nauk*, vol. 2, 599-602. As a student of Euler, Rumovskii also viewed skeptically Newton's ideas regarding gravitation, but he seemed to look even more askance at Lomonosov's alternative theorizing.

reputation to that of Wolff and Euler was a wonderful vehicle for firmly establishing his personal status and honor, and somewhat less so that of the natural philosopher, in Russian society. It also had the effect of solidifying his image as a pioneering Russian scientist, which succeeding generations would reshape, while leaving intact the fundamentals, into a likeness that they found most appropriate for their times. The mythology of Lomonosov developing out of this process became ever more central to Russian cultural and scientific pretensions.

"The Young Lomonosov on the Path to Moscow"
Painting by N. I. Kisliakov, 1948

Chapter 2

RUSSIA'S
"OWN PLATOS AND QUICK-WITTED NEWTONS":[1]
INVENTING THE SCIENTIST

*I*urii Lotman, in his analysis of "everyday behavior" in the eighteenth century,[2] argued that the seeking after a stylized ideal, the "self-assessment" assumed by the subject, also to an extent governed his future actions and how they would be "received." His first point seems a truism; the reception of the image is the intriguing element. Lotman believed that this selection of idealized figures "gave rise to anecdotal epics which were built up on the principle of accumulation." This was a behavior that "was in principle open-ended: it could be infinitely expanded, being enriched by ever more incidents." The biographies of Lomonosov written by his younger contemporaries in the last three decades of the eighteenth century strongly testify to the notion of "accumulation." These biographical accounts also expose the instrumental value of Lomonosov's self-fashioning as it molded and unmistakably delimited how Lomonosov's life was recorded by deferential observers.

His early biographers would work within the pattern of heroic tales. However, in providing trenchant critiques of Lomonosov's place in the history of science, accompanied by their thoughtful reading of his scientific papers and their knowledge of

---

1   Lomonosov, 1747 (Lomonosov, *PSS*, vol. 8, 206).

2   Lotman, "Poetics of Everyday Behavior," 241-242.

contemporary trends in natural philosophy, they also launched the genre, admittedly long embryonic, of scientific biography in Russia. The rich admixture of analysis, "fact," and anecdote, always difficult to separate, provided ample room for the development and evolution of a mythology around Lomonosov's life.

The accumulation of publications, counting only the number of studies devoted to Lomonosov, commenced apace as early as in the first few years following his death. The value of these initial works in adding to the building of the Lomonosov legend was negligible, however: the ones composed in Russia were largely ignored, and those published abroad made little impression in Russia at the time, though they were later the subjects of much attention. That Lomonosov, apparently surrounded by enemies at the Academy of Sciences, did not receive a proper encomium following his death on 4 April 1765 became part of the legend that suffused his name.

Although he was not the recipient of a eulogy, comparable to those most famously conferred by Fontenelle, and later Condorcet, of the Paris Academy on selected worthies,[3] he was hardly excised from sight. Nicolas Le Clerc, a French physician newly elected to an honorary membership in the Academy of Sciences, delivered an address of acceptance (15 April 1765) that included a sizeable passage venerating Lomonosov's services to Russia in the field of literature. However, he omitted mention of Lomonosov's work in the sciences. Le Clerc's oration was met without enthusiasm by the Academy members,[4] was not made available for distribution, and

---

[3]   There is no adequate study of the eulogies offered at the St. Petersburg Academy of Sciences, but for limited comparative flavor, please see C. B. Paul, *Science and Immortality: The Eloges of the Paris Academy of Sciences (1699-1791)* (Berkeley, 1980), 1-27, 133-55 (which contains three examples of eulogies); Dorinda Outram, "The Language of Natural Power: the 'Eloges' of Georges Cuvier and the Public Language of Nineteenth-Century Science," *History of Science* 16, no. 33 (September 1978): 153-78. The eulogies pronounced at the Paris Academy were widely disseminated in European scientific circles.

[4]   *Protokoly zasedanii Konferentsii Imperatorskoi Akademii nauk s 1725 po 1803 goda*, vol. 2, 1744-1770 (St. Petersburg, 1899), 536-37.

was consigned to the archives.[5] Jacob von Staehlin, Lomonosov's longtime colleague at the Academy of Sciences, prepared a eulogy for him, but neither delivered nor published it at the time.[6] Whether it was withdrawn because of the enmity of members of the Academy who did not wish to honor Lomonosov, or because of some rancor towards Staehlin, or for any number of other reasons, is unknown. Staehlin's eulogy served as the basis for a much more substantial essay on Lomonosov that he wrote in the 1780s, which will be discussed shortly.

Two brief tracts on Lomonosov were also composed abroad soon after his death. That these "foreign" studies were also both published has given some sustenance to those who accept the notion that Lomonosov was surrounded by enemies at the Academy who prevented the bestowal upon him of his rightful rewards. In 1765 Andrei Shuvalov (1743-89), a minor poet, relative of Ivan Shuvalov, and distant acquaintance of Lomonosov, then living in Paris, penned "Ode sur la mort de Monsieur Lomonosov de L'Académie des sciences de St. Petersbourg."[7] It is Shuvalov's introduction to his ode

---

[5]   A century later it was discovered by Pekarskii, who subsequently published the section concerning Lomonosov in "O rechi v pamiat' Lomonosova, proiznesennoi v Akademii nauk doktorom Le-Klerkom," *Zapiski Imperatorskoi Akademii nauk* 10, book 2 (1867): 178-81. For discussions of Le Clerc's speech and the issue of the insult to Lomonosov's memory by neglecting not only to provide a proper eulogy, but also to publish the one on hand, see Pavlova, *Lomonosov v vospominaniiakh,* 5, 21; Radovskii, *Lomonosov i Akademiia nauk,* 222-23; V. A. Somov, "N.-G. Leklerk o M. V. Lomonosove," in *Lomonosov: sbornik statei,* vol. 8 (Leningrad, 1983), 97-105. Somov holds that Le Clerc's incautious approach to Peter I and to Russia's past, in addition to the general hostility toward Lomonosov on the part of some of the academicians, caused the overall piece to fail to gain approval.

[6]   M. P. Pogodin was the first to publish Staehlin's "A Summary of Lomonosov's Eulogy" ("Konspekt pokhval'nogo slova Lomonosovu"), printing it in *Moskvitianin,* no. 3 (1853): 22-25. At the head of his piece Staehlin wrote that it was those who scorned Lomonosov's legacy who had prevented it from being read at the Academy. There is no evidence to solidly support Stachlin's contention.

[7]   In this publication Shuvalov also included his French translation of Lomonosov's *Utrennee razmyshlennie o bozhiem velichestve.* These works were first published in Russia, although only in French, in 1865. See Kunik,

that is especially relevant for this study.[8] Although much of it is spent hailing Lomonosov's literary and linguistic accomplishments, with which Shuvalov was very familiar, he recounts the epic journey, for the first time in print, of Lomonosov's travels from the far north of Russia, where "early in life he exhibited a love for the sciences," to Moscow, and then to Marburg, and then lastly Freiberg. Shuvalov's noting of Freiberg, where Lomonosov "studied metallurgy and mining" was, with its nod toward the need to master the practical sciences, a foreshadowing of a future theme in studies of Lomonosov: the emphasis on the real, not simply theoretical, benefits that science and scientists could bring to Russia. For Shuvalov, this was also a far more accessible area of the sciences than chemistry and physics.

Throughout Shuvalov's introduction, Lomonosov's "energy," "talent," and unspecified striving for "knowledge" and "new ideas" appear in the background. Shuvalov counted Lomonosov particularly lucky to have been able to go abroad, where it was "possible to study much that was new, and also where he was fortunate enough to learn from the famous Wolff." Eschewing any perusal of Lomonosov's papers or activities in "science," except for a citation from the *Letter on the Usefulness of Glass*, Shuvalov observed that he was named professor of chemistry at the Academy of Sciences by Elizabeth I and was the "first scientist in Russia." A dry listing of selected public achievements and published writings, along with titles or ranks held, was de rigueur in eighteenth-century biographies. Attempts to address the content of the subject's working or intellectual life intensified toward the end of

---

Sbornik materialov, part 1, 201-10. Shuvalov dispatched the publication to Voltaire, who responded with a letter to Aleksandr Vorontsov commenting that he "would always remember the beautiful verses [Lomonosov's], that he [Shuvalov] translated into our language." *Arkhiv kniazia Vorontsova*, vol. 5 (Moscow, 1872), 455. See also D. F. Kobeko, "Uchenik Vol'tera graf Andrei Petrovich Shuvalov," *Russkii arkhiv*, book 3 (1881): 250, 252, 257-58.

[8]   Shuvalov's introduction was not translated into Russian until 1936. See P. N. Berkov, *Lomonosov i literaturnaia polemika ego vremeni* (Moscow-Leningrad, 1936), 277-79.

the century.[9] Shuvalov's short biography is better studied as part of the myth of Lomonosov as the Russian Malherbe or Pindar than as integral to representations of him as the father of Russian science.[10] He was interested in Lomonosov as a natural philosopher solely as that image interacted with the remarkable tale of a youth from the periphery rising to a commanding position in the Academy of Sciences.

In 1768 a biographical review of Russian writers, "Nachricht von einigen russischen Schriftstellern, nebst einem kurzen Berichte vom russischen Theater," appeared in the Leipzig journal *Neue Bibliothek der schönen Wissenschaften und der freyen Kunste*, with a brief entry on Lomonosov.[11] Though by definition interested in

---

[9]   W. Gareth Jones, "Biography in Eighteenth-Century Russia," *Oxford Slavonic Papers* 22 (1989): 58; Peter J. Korshin "The Development of Intellectual Biography in the Eighteenth Century," *Journal of English and Germanic Philology* 73, no. 4 (October 1974): 513-23.

[10]   In his *Epistle on Versification* (*Epistola o stikhotvorstve*, 1747) Aleksandr Sumarokov wrote of Lomonosov, "He is the Malherbe of our Lands, he is like Pindar." A. P. Sumarokov, *Izbrannye proizvedenniia* (Leningrad, 1957), 125. Sumarokov was not proposing an aesthetic identity between Lomonosov's poetry and that of either Pindar's or Malherbe's; rather he was ascribing to Lomonosov a correspondingly pioneer status in the establishment of Russian letters. Sumarokov's epigrammatic phrase swiftly became a cliché in studies of Lomonosov, and of eighteenth-century Russian literature. In light of Sumarokov's later enmity towards Lomonosov, it is a poetic legacy whose irony he would presumably not have appreciated. See also G. A. Gukovskii, *Russkaia poeziia XVIII veka* (Leningrad, 1927), 32-33; Pekarskii, *Istoriia Akademii nauk*, vol. 2, 133-34; Reyfman, *Trediakovsky*, 59-61, 91.

[11]   This was first translated into Russian and published in 1867. See P. A. Efremov, ed., *Materialy dlia istorii russkoi literatury* (St. Petersburg, 1867), 131-33. Gareth Jones contends that this ariticle appeared at the behest of Russian authorities in specific response to the Chappe d'Auteroche's *Voyage en Sibérie* (1768), which he quite accurately described as a "poisonous calumny" against Russia's apparent political, social, and cultural backwardness. W. Gareth Jones, "The Image of the Eighteenth-Century Russian Author," in *Russia in the Age of the Enlightenment: Essays for Isabel de Madariaga*, ed. Roger Bartlett and Janet Hartley (New York: St. Martin's Press, 1990), 63-64. Marcus C. Levitt, in "An Antidote to Nervous Juice: Catherine the Great's Debate with Chappe d'Autoeroche over Russian Culture," *Eighteenth-Century Studies* 32, no. 1 (1998): 56, 62 disputes Jones's view on the inspiration for

its subject's literary activities, this piece, translated into French in 1771, long served, with Shuvalov's work, as the principal source on Lomonosov's life for foreign audiences. The German essay expended considerable space in attempting to flesh out the relationship between Lomonosov and the poet and dramatist Aleksandr Sumarokov. Though their relationships with each other and with Vasilii Trediakovsky were fundamental to the formation of Russian literary reputations, they had no perceptible influence on Lomonosov's perceived scientific legacy.[12] The anonymous author of the German article—referred to in the work as simply a "Russian traveler"—credited Shuvalov's ode as being an important source for his study. However, Lomonosov's journey to knowledge, with a commensurate focus on his early diligence and natural gifts, which only grew as he aged, and were thematically crucial to Shuvalov's account, were absent in the *Neue Bibliothek* piece. Tales depicting the Homeric wanderings of the young Russian may have had less resonance for a foreign audience.

The most thoughtful early studies of Lomonosov that deal at some length with his science are Mikhail Murav'ev's "Contributions of Lomonosov to Learning" ("Zaslugi Lomonosova v uchenosti") and Aleksandr Radishchev's "Discourse on Lomonosov" ("Slovo

---

the German article, though he fails to offer an alternative explanation. The Chappe d'Autoeroche, a French astronomer, compiled his three-volume book—a compendium quite likely sponsored by anti-Russian circles in the French government—following a trip to Russia he made in 1761 in order to observe the passage of Venus before the Sun. In 1770 Catherine the Great authored, in French, a detailed rebuttal of the Chappe d'Autoeroche's work. Entitled *Antidote*, it was clearly aimed at those in Western Europe who would defame Russia.

[12] Among the better guides to the massive topic of Lomonosov's often bitter literary relationships with Sumarokov and Trediakovsky are Berkov, *Lomonosov i polemika*; 92-272; Reyfman, *Trediakovsky*, 49-69; Serman, *Lomonosov*, 188-208; B. A. Uspenskii, *Vokrug Trediakovskogo: trudy po istorii russkogo iazyka i russkoi kul'tury* (Moscow, 2008); Zhivov, *Iazyk i kul'tura v Rossii XVIII veka*; and idem, "Pervye russkie literaturnye biografii." Reyfman and Zhivov approach the posthumous creation of a "cult" (Zhivov's term) of Lomonosov and the consequent downward trajectory of Trediakovskii's and Sumarokov's authorial standings in refreshingly provocative ways.

o Lomonosove"). Radishchev's and Murav'ev's works, both fascinating attempts to assess Lomonosov's overall stature as a natural philosopher, are not only singular contributions to the genre of scientific memoir in their own right, but sharply divergent responses to the images of Lomonosov as the pioneer Russian scientist that were firmly situated in the cultural dialogue of the day. These representations were not only implanted by the processes already discussed, but by biographies written by figures of renown themselves, who added crucial details necessary to the continuing growth of Lomonosov's fame as Russia's first "man of science."[13]

Nikolai Novikov's "Lomonosov, Mikhailo Vasil'evich," published as part of his *An Attempt at a Historical Dictionary of Russian Writers* (*Opyt istoricheskogo slovaria o rossiiskikh pisateliakh*, 1772); Jacob von Staehlin's "Traits and Anecdotes for a Biography of Lomonosov, Taken from His Own Words by Staehlin" ("Cherty i anekdoty dlia biografii Lomonosova, vziatye s ego sobstvennykh slov Shtelenym," 1783)—the fullest of his several pieces on Lomonosov, and the most influential of the eighteenth-century studies; and Mikhail Verevkin's "Life of the Late Mikhail Vasil'evich Lomonosov" ("Zhizn' pokoinogo Mikhaila Vasil'evicha Lomonosova," 1784), served until the middle of the next century as the essential sources on Lomonosov's life.[14] Considering the

---

13   Read consecutively, Steven Shapin's *A Social History of Truth* and *The Scientific Life: A Moral History of a Late Modern Vocation* (Chicago: University of Chicago Press, 2008), reconstruct how natural philosophers became "men of science" (and later scientists).

14   D. S. Babkin, G. E. Pavlova, and especially Irina Reyfman have each examined these biographies. Babkin was most interested in the intricacies of their composition, while Reyfman's work is concerned with how they created and exaggerated Lomonosov's literary reputation. Galina Pavlova suggested their potential value as critiques of Lomonosov's science, but did not pursue this line of inquiry. See Babkin, "Biografii M. V. Lomonosova sostavlennye ego sovremennikami," in *Lomonosov: sbornik statei*, vol. 2 (Moscow-Leningrad, 1946), 5-70 and Pavlova, *Lomonosov v vospominaniiakh*, 3-13; Reyfman, *Trediakovsky*, 8-10, 260, 273. Martynov, *Lomonosov: glazami sovremennikov*, 320-43, and Smagina, *Kniaginia i uchenyi*, 227-334, include useful publication details on contemporary biographies of Lomonosov.

underdeveloped nature of memoir writing in the eighteenth century, and not only in Russia, this amount of attention was remarkable.[15] To avoid unnecessary repetition, as well as to illustrate how these essays worked within, and broadened, an existing narrative of Lomonosov's life, Staehlin's and Verevkin's texts will be discussed together. Because it is thinner in all respects than its two successors, Novikov's biography of Lomonosov will be looked at separately.

As part of his long efforts to project the notion that Russia possessed its own rich literary traditions, in 1772 the publisher and writer Nikolai Novikov (1744-1818) published *An Attempt at a Historical Dictionary of Russian Writers*.[16] Novikov's account of

---

V. P. Lystsov provides a broad survey of early literature devoted to Lomonosov in his *M. V. Lomonosov v russkoi istoriografii 1750-1850-kh godov* (Voronezh, 1983), 3-69. Lystsov's book, however, is terribly undermined by the author's crude social-political categorizations and analyses.

[15] The standard reference directory to published materials for eighteenth-century Russia, and an invaluable source of disparate biographical information, remains the *Svodnyi katalog*. It merits noting, however, that its classification system does not contain a separate biography category, perhaps reflecting, as Gareth Jones remarked, the "absence of biography as a distinct genre" in eighteenth-century Russia. It may also reveal that the editors of the *Svodnyi katalog* did not perceive the eighteenth-century emergence of biographical writing. Gareth Jones's article concludes with a seemingly complete listing of biographies published in eighteenth-century Russia. See Jones, "Biography," 58, 71-79. For a compilation of unpublished eighteenth-century Russian memoirs, see A. G. Tartakovskii, *Russkaia memuaristika XVIII-pervoi poloviny XIX v. Ot rukopisi k knige* (Moscow, 1991), 244-70.

[16] N. I. Novkov, ed., *Opyt istoricheskago slovaria o rossiiskikh pisateliakh* (St. Petersburg, 1772; reprint, Leningrad, 1987), 119-30. Novikov stated in the preface to his dictionary that his "incentive" to compose it derived from what he perceived as certain inequities and biases in the aforementioned Leipzig contribution on Russian writers. Novikov welcomed the article's appearance — it was after all the first of its type devoted to Russian literature — but he also implied that the anonymous "Russian traveler" who drafted it displayed an inadequate understanding of the expansive breath of Russian culture. I. F. Martynov persuasively submitted that Novikov was driven to write his dictionary not only by deficiencies in the German article, but also as part of a continuing Russian reproach to the Chappe d'Autoeroche's opprobrium. See I. F. Martynov, "'Opyt istoricheskogo slovaria o rossiiskikh pisateliakh' N. I. Novikova i literaturnaia polemika 60-70-kh godov XVIII

Lomonosov in the *Dictionary* was reissued in the 1778 three-volume edition of Lomonosov's papers published by Moscow University.[17] Due to the immense prestige in Russian cultural history of both subject and author, it has been reprinted numerous times in the two centuries since it was first released. This examination of the essay will concentrate on its addressing of Lomonosov's scientific legacy. Novikov had no special training in natural philosophy, nor was it likely that he ever met Lomonosov. It has been suggested, though, that Novikov was at least familiar with Lomonosov's oft-cited *The First Principles of Metallurgy or Mining* (*Pervyie osnovaniia metallurgii ili rudnykh del*, 1763).[18] He was, moreover, acquainted with personages close to Lomonosov and had a sure feel for the intellectual and cultural life of the time.

Much of Novikov's "Lomonosov" is spent on a recitation of the titles (adjunct and then professor of chemistry at the Academy of Sciences) and ranks (Collegiate and then State Councilor) held by his subject. Various important dates are assigned to events in his life (Novikov is rarely more than a year off), and a partial list of his works published at home and those translated for foreign consumption is included. Novikov also included the Russian and Latin inscription that Staehlin had composed for the monument erected by Lomonosov's patron, Mikhail Vorontsov, over

---

veka," *Russkaia literatura*, no. 3 (1968): 187. Interestingly, Lomonosov served, in the Chappe d'Autoeroche's opinion, as an exception to the bleakness of Russian attainments in the arts and sciences. On Novikov's *Dictionary* in the context of enlightenment in Russia, particularly useful is Colum Leckey, "What is *Prosveshchenie*? Nikolai Novikov's *Historical Dictionary of Russian Writers* Revisited," *Russian History* 37 (2010): 360-77.

[17] Lomonosov, *Sobranie raznykh sochinenii v stikhakh i v proze*, vol. 1 (Moscow, 1778). The practice of including a brief biography of the author with multi-volume editions of their works was established, most probably first in England, in the seventeenth century. See Hall, *Isaac Newton*, 3.

[18] K. V. Ostrovitianov, ed., *Istoriia Akademii nauk SSSR*, vol. 1 (1724-1803) (Moscow-Leningrad, 1958), 232. According to the authors, among the readers of Lomonosov's mining manual were Radishchev and Diderot. For its publication figures in the eighteenth century, see *Svodnyi katalog*, vol. 2, 172.

Lomonosov's grave.[19] This inscription parallels Novikov's listing of titles and ranks.

Novikov's memoir of Lomonosov fits the then-conventional mold of listing great deeds in later life that were necessarily preceded by a precocious childhood and adolescence. Though coming from the far provinces, Lomonosov, the son of a fisherman, was able to read and write in his youth. Indeed, he "showed an early inclination toward the reading of books." Novikov remarked that Lomonosov, in adolescence, "by good fortune got ahold of the *Psalter* written by Simeon Polotskii, and having read it over many times, became enamored of poetry and wanted to know where he could study prosody."[20] Absent in Lomonosov's writings, however, is any mention of this famed prosodic guide. Knowledge of the *Psalter* by Lomonosov in such a place and at such a relatively early age presumably signaled to the reader, as it did to Novikov, that Lomonosov was an extraordinary child. Novikov continued the story with Lomonosov learning that the art of versification could be acquired at the Slavo-Greco-Latin Academy in Moscow, journeying there, and "with great diligence" studying "Latin and Greek, rhetoric and versification."

Novikov's text follows Lomonosov's travels from Moscow to the Academy of Sciences in St. Petersburg, and from there to Marburg for studies with the "renowned Wolff." In Marburg, he was "trained in chemistry and in the related sciences." Lomonosov's year in Freiberg with the chemist Johann Henkel was devoted to the study of "mineralogy and mining."[21] Novikov does not provide any details of the science Lomonosov learned along the way, but

---

[19]   Novikov, "Lomonosov," 123-26.

[20]   Polotskii's *Psalter*, composed in 1680, was long the premier handbook to prosody in Russia. In his generous reconstruction of Lomonosov's library G. M. Korovin speculates, but is unable to confirm, that Lomonosov must have been familiar with Polotskii's manual. G. M. Korovin, *Biblioteka Lomonosova: materialy dlia kharakteristiki literatury, ispol'zovannoi Lomonosovym v ego trudakh, i katalog ego lichnoi biblioteki* (Moscow-Leningrad, 1961), 6-7, 310-11.

[21]   Novikov, "Lomonosov," 119-21.

the image of the youthful Lomonosov, emerging from such an inhospitable locale as the far north of Russia, going abroad for studies with someone of Wolff's stature, all for the pursuit of the sciences, was conveyed as astounding.

In assessing Lomonosov's twenty-five years of service at the Academy of Sciences, Novikov wrote that "strong was his striving toward the sciences and toward all useful knowledge." He always worked "to overcome all obstacles and was rewarded with great success." Novikov was impressed by Lomonosov's command of languages, so important for the sciences, which, with exaggeration, he listed as including abilities, to varying degrees of fluency, in German, Latin, French, and Greek (Lomonosov did begin the study of Greek, but quickly gave it up). In addition, he stressed that for the time, Lomonosov's knowledge of the essence of Russian, along with his enrichment of it, and presumably of its scientific vocabulary, was quite unrivaled. "Trained in all the philosophical and literary sciences, in chemistry, with its different parts," Lomonosov "was especially accomplished in experimental physics, which he translated into the Russian language [Novikov's reference is to the *Wolffian Experimental Physics*], in mechanics, and in the history of our country."[22]

Though not interested, or perhaps even aware, of his theoretical work at the Academy of Sciences, Novikov did cite Lomonosov's work on the mineralogy cabinet of the *Kuntskamera*, and accentuated his work on mosaics. Apropos a mosaic honoring Peter the Great (most probably *The Battle of Poltava*), "the likes of which has not been repeated," he wrote that Lomonosov "finished the work with Russian masters and materials and without any sort of help from foreigners."[23] More interestingly for our purposes,

---

[22]   Ibid., 127.

[23]   Ibid., 122. Lomonosov worked sporadically for more than ten years on the creation of mosaic art; it was among his activities that were most widely commented on in the eighteenth century. Indeed, when Catherine II visited him in 1764, a contemporary report emphasized her inspection of Lomonosov's mosaics for a planned monument to Peter I. See *Sanktpeterburgskie vedomosti*, no. 48 (15 June 1764). For an authoritative

he commented in the last paragraph that Lomonosov had "been engaged in correspondence with many scientists in Europe."[24]

Novikov's earlier reference to Wolff"s connections to Lomonosov, in addition to Lomonosov's status as a "colleague" of Euler's—of which his readers were aware—demonstrated that Lomonosov was a figure comparable to heralded Western scientists. To Novikov he was the only scientist of such august stature Russia had yet produced. Novikov attached to his biography of Lomonosov a short poem (six lines of verse) written by Nikolai Popovskii, a former student of Lomonosov's best known for his translation of Alexander Pope's *Essay on Man*, to honor the late scientist. The last two lines read: "Otkryl natury khram bogatym slovom Rossov; Primer ikh ostroty v naukakh Lomonosov" ("He opened nature's temple with the rich language of the Russians; Lomonosov is an example of their keenness in the sciences").[25]

---

discussion of Lomonosov's mosaics, see Makarov, *Khudozhestvennoe nasledie Lomonosova*, 7-126. Jacob von Staehlin penned an interesting description of the mosaic arts in Russia, and Lomonosov's role in their development. Staehlin's notes were not uncovered and published until Makarov's work on them in 1950 (ibid., 279-98).

[24]   Novikov, "Lomonosov," 130.

[25]   Ibid., 129. The full verse inscription, "Nadpis' k portretu M. V. Lomonosova," was initially printed under Lomonosov's portrait in the first volume of the 1757-59 two-volume edition of his collected works (despite the dating, the first volume actually came out in 1758, with the second issued in 1765). There has been some confusion as to the identity of the author of this verse. In his biography of Lomonosov, Novikov identified Nikolai Popovskii as the poem's creator. At the conclusion of his section on Ivan Shuvalov in this same *Dictionary of Russian Writers*, however, he designated Shuvalov as the author (ibid., 249). For knowledgeable discussions of this question that nevertheless still leave the inscription's provenance ambiguous, see Kobeko, "Uchenik Vol'tera," 257-58; and Pekarskii, *Istoriia Akademii nauk*, vol. 2, 625-26. A letter of 7 November 1758 from Sumarokov to Ivan Shuvalov, located in G. P. Makogonenko, ed., *Pis'ma russkikh pisatelei XVIII veka* (Leningrad, 1980), 84, 191, in which Sumarokov indicates, albeit somewhat obliquely, that Popovskii composed the work, appears to further substantiate his authorship. Although Pekarskii printed this same letter, he draws no definitive inferences from it.

Novikov never made a distinction between Lomonosov's interests in science and his involvement in literature. At this early point in the development of memoir writing in Russia the substance, be it scientific or literary, of the subject's work was not presented in such a way that would anchor it to the progression of the life outlined in the biography. Without an attempt to consider thoughtfully the course of Lomonosov's life beyond his childhood years, we are left with his journey to enlightenment with no real study of his intellectual evolution en route or after his "arrival." Absent from Novikov's biography was the plethora of detail and anecdote that would enliven the framework of Lomonosov's life in Staehlin's and Verevkin's essays. Within Novikov's "Lomonosov" it is the quest for knowledge and the way stations visited along the way, where much of that knowledge was gained, that made up the image of Lomonosov. The *Psalter*, Christian Wolff, chemistry, and the Academy of Sciences were the important signifiers to Novikov's audience.

It has become a truism in the literature that from the time of the so-called Scientific Revolution until quite recently, reminiscences of natural philosophers and collected anecdotes about them largely determined the way various interested publics received them and the scientific community they represented.[26] Spectacular scientific achievement was intimately linked to the possession by the natural philosopher of a level of virtue and/or heroism measured in epic proportions. This was fundamental to the incipient scientific community's efforts to establish a degree of legitimization in society. Images of Robert Boyle were long dependent on highly idealized accounts of his life composed soon after his death that remain influential sources to this day. In examining early representations of Boyle (perhaps most influential was Gilbert Burnet's sermon), Michael Hunter and Steven Shapin have persuasively delineated how his apparently irreproachable life was presented as a model for what it meant to be a scientist in seventeenth-century England.[27]

---

26   See preface to Shortland and Yeo, *Telling Lives in Science*.

27   Michael Hunter, "Robert Boyle and the Dilemma of Biography in the Age

If the requirement to emphasize the virtues of Boyle had a rather unhealthy effect on attempts to compile a more "objective" account of his life, in the case of Isaac Newton, who had supplanted Boyle as an ideal type in the eighteenth century, early elevated portraits of Newton (John Conduitt's anecdotes, Fontenelle's eulogy, and Thomas Birch's article were among the most important) fused with the inviolability of his scientific work to structure a nearly unassailable heroic image. Studies by Rupert Hall and Rebekah Higgitt on foundational biographies of Newton soundly attest to how attempts in the nineteenth century to present a more balanced view of Newton's life were seen as akin to a violent attack on a national symbol.[28] In Russia, Lomonosov came to be portrayed as a symbol as important to his people—if not yet to the world, though that proclamation would come with Soviet-era assertions— as Newton was to his. Lomonosov's self-presentation provided the initial narrative. Building on Novikov's biography, Verevkin and Staehlin served as Lomonosov's Conduitt, Fontenelle, and Birch.

A translator and dramatist associated with Moscow University, Mikhail Verevkin (1732-95), director of the gymnasium in Kazan', corresponding member of the Academy of Sciences and member of the Imperial Russian Academy (founded in 1783 and dedicated to the study of the Russian language and literature), wrote his biography of Lomonosov, "Life of the Late Mikhail Vasil'evich Lomonosov," for the 1784-87 edition of Lomonosov's collected works.[29] Whether or not Lomonosov and Verevkin ever met has not

---

of Scientific Revolution," in ibid., 115-117, 133-34; and idem, *Robert Boyle by Himself and his Friends*, XI-C (this collection also includes the texts of several autobiographical and contemporary biographical accounts); Shapin, *Social History of Truth*, 130-92. For an interesting example of constructions of the representative scientist in the more modern era, see Geoffrey Cantor, "The Scientist as Hero: Public Images of Michael Faraday," in Shortland and Yeo, *Telling Lives in Science*, 171-91.

[28] Hall, *Isaac Newton*, 180-92; Higgitt, *Recreating Newton*, 19-127. See also Haynes, *From Faust to Strangelove*, 50-65; Yeo, "Images of Newton," 270-79.

[29] *Polnoe sobranie sochinenii Lomonosova*, vol. 1, 1784, III-XVIII. This biography, which has been reprinted many times, can also be found in Pavlova, *Lomonosov v vospominaniiakh*, 42-51. There is a paucity of biographical

been firmly established. Verevkin was, however, closely associated for a time with Moscow University's first curator, Ivan Shuvalov, who, it can be supposed, was a valuable source for information on Lomonosov. Verevkin's "Life of Lomonosov" was, along with Novikov's composition, the main published biographical work on Lomonosov within Russia until well into the next century.

Jacob von Staehlin (1709-85), who came to the St. Petersburg Academy of Sciences in 1735 from the University of Leipzig as a poetry adjunct, with a mandate to teach his subject in the "German style," and to instruct the students in "eloquence and versification, as well as to undertake the production of illuminations and fireworks and so forth, and ... future exercise in the sciences and arts,"[30] spent the next half-century in Russia (he was named professor of eloquence and poetry and a full member of the Academy of Sciences in 1737) and enjoyed a career of remarkable versatility.[31] Composer of odes, translator, editor of Academy publications (notably *Sanktpeterburgskie vedomosti*), tutor to the future Emperor Peter III, longtime director of the arts departments at the Academy of Sciences, and producer of many of the spectacles so much

information on Verevkin. The most comprehensive remains that issued in *Russkii biograficheskii slovar'*, vol. 3A (Petrograd, 1916; reprint, New York, 1991), 582-85 (entry by V. Korkeakova), though it is still inadequate. See also Iu. V. Stennik, "Verevkin, Mikhail Ivanovich," in *Slovar' russkikh pisatelei XVIII veka*, no. 1 (Leningrad, 1988), 148-50.

30　*Zapiski Iakoba Shtelina ob iziashchnykh iskusstvakh v Rossii*, vol 1, comp. K. V. Malinovskii (Moscow, 1990), 8.

31　Staehlin's activities have not yet received the attention they merit, but Pekarskii's study of Staehlin still has much to recommend it. Pekarskii, *Istoriia Akademii nauk*, vol. 1, 538-67. See also the biographical sketch attached to *Zapiski Iakoba Shtelina*, 7-32, and Klaus Harer, "Ia. Shtelin i M. V. Lomonosov. Novyi vzgliad na ikh vzaimootnosheniia po arkhivnym istochnikam," in *Lomonosov: sbornik statei*, vol. 10, 180-92. James Cracraft examines Staehlin's views on Russian visual arts in his *Petrine Revolution in Russian Imagery*, 204-08, 216-17. Cracraft also added a brief outline of Staehlin's life (ibid., 203-04). Francine-Dominique Liechtenhan, "Jacob von Stählin, académicien et courtesan," *Cahiers du monde russe*, 43, 2-3 (2002): 321-32, highlights Staehlin's frequently dismissive views of his colleagues at the Academy, along with his efforts to secure ever higher status in official St. Petersburg.

a part of court ceremonies, Staehlin was at the center of Russian intellectual life for five decades. He was also, for some twenty-five years, a close colleague, and occasional collaborator, of Lomonosov at the Academy of Sciences.[32]

Though Staehlin's associations with Lomonosov were per-haps closest in the area of designing pyrotechnical displays and in the writing of ceremonial odes,[33] they were also both very involved

---

[32] Much time has been devoted to investigating whether Staehlin was a "friend" or "enemy" of Lomonosov. It seems clear that the need to assign Staehlin a role as one of Lomonosov's "persecutors" fits comfortably into a leitmotif in the literature that supports beliefs that enemies prevented Lomonosov from completing his work and serving the people. Pekarskii provided slender evidence, in the form of a letter from Lomonosov to the Chancellery of the Academy of Sciences, a document interpreted very loosely in later accounts, which indicates that in the 1750s Lomonosov and Staehlin quarreled over the arrangement of fireworks (Pekarskii, *Istoriia Akademii nauk*, vol. 1, 547-48). See also Lomonosov, *PSS*, vol. 10, 350. Makarov, referring to records used by Staehlin's descendant Karl Staehlin in a 1926 publication, showed that in 1763 Staehlin and some of his colleagues in the administration of the Academy of Sciences might have tried to oust Lomonosov from the Academy (Makarov, *Khudozhestvennoe nasledie Lomonosova*, 282). Notwithstanding their sharply different arguments and objectives, Galina Pavlova and Irina Reyfman each saw it as consequential that Staehlin not be labeled a "friend" of Lomonosov (Pavlova, *Lomonosov v vospominaniiakh*, 8-9; Reyfman, *Trediakovsky*, 260). For a searching examination of the question that especially contests Makarov's explication of Staehlin and Lomonosov's relationship, please consult *Zapiski Iakoba Shtelina*, 111-116.

[33] Lomonosov's extensive involvement from 1747 to 1755 on the production of fireworks, illuminations, and the accompanying literary adornments—mainly poetic inscriptions for the pyrotechnics—is testified to by his voluminous oeuvre in this area. See Lomonosov, *PSS*, vol. 8, 194-579, 933-1043. Much of Lomonosov's work on court spectacles was either in close collaboration with or under the supervision of Staehlin. See also Galina Pavlova, "Proekty illiuminatsii Lomonosova," in *Lomonosov: sbornik statei*, vol. 4 (Moscow-Leningrad, 1960), 219-37; Barbara Widenor Maggs, "Firework Art and Literature: Eighteenth-Century Pyrotechnical Tradition in Russia and Western Europe," *Slavonic and East European Review* 54, no. 1 (January 1976): 28-29, 34; Hans Röhling, "Illustrated Publications in Fireworks and Illuminations in Eighteenth-Century Russia," in *Russian and the West in the Eighteenth Century*, ed. A. G. Cross (Newtonville, MA: Oriental Research Partners, 1983), 95-6; and D. D. Zelov, *Ofitsial'nye svetskie prazdniki kak iavlenie russkoi kul'tury kontsa XVII—pervoi poloviny XVIII veka*, 2nd ed.

in administrative affairs at the Academy of Sciences.[34] Written from a better vantage point than anyone else who wrote about Lomonosov, Staehlin's "Traits and Anecdotes for a Biography of Lomonosov, Taken from His Own Words by Staehlin,"[35] is a rich combination of personal recollections that contrast starkly with the dry listing of dates and works which characterizes Novikov's "Lomonosov." It is also a comprehensive introduction to Lomonosov's early life and journeying to Moscow, Petersburg, Marburg, and around the various German lands for "knowledge."

Both Verevkin's and Staehlin's entries focus most of their attention on the years before Lomonosov became a professor at the Academy of Sciences. These biographers were intent on explaining to their readers how this figure of such humble provenance came to be counted among the foremost personages of his time. It bears repeating that this was emblematic of eighteenth-century scientific memoirs. In substance, Verevkin's and Staehlin's analyses of Lo-

---

(Moscow, 2010), 239-51. G. S. Smith convincingly argues that Lomonosov's early poetic output was under the tutelage of German scholars, most principally Staehlin, at the Academy of Sciences. See his article "The Most Proximate West: Russian Poets and the German Academicians, 1728-41," in *Russia and the World of the Eighteenth Century*, ed. R. P. Bartlett, A. G. Cross, and Karen Rasmussen (Columbus, OH: Slavica Publishers, 1988), 366-67. L. V. Pumpianskii's "Lomonosov i nemetskaia shkola razuma," *XVIII vek* 14 (1983): 3-44, on the other hand, attempts to demonstrate that Lomonosov's fundamental poetic assumptions were formulated quite independently of any Germanic influences.

[34] Due to Schumacher's increasing infirmities, the administration of the Academy of Sciences was reorganized early in 1757. Lomonosov and Taubert, followed soon after by Staehlin, were named to its governing chancellery. For the next several years, as is demonstrated in Lomonosov, *PSS*, vol. 10, 194-316, 649-737, passim, Lomonosov and Staehlin were immersed in managing the Academy (these documents relate to their work in the chancellery, and are only a partial record of it). See also Radovskii, *Lomonosov i Akademiia nauk*, 179-221. Radovskii, perhaps not unexpectedly in a monograph so determined to extol Lomonosov's role as the preeminent organizer of eighteenth-century Russian science, pays less heed to the value of Staehlin's service.

[35] Staehlin's memoir of Lomonosov was first published in *Moskvitianin*, no. 1 (1850): 1-14. Prior to this it was circulated widely in manuscript.

monosov's work at the Academy of Sciences is hardly fuller than Novikov's. However, the added information on his early intellectual development (through his time abroad) and the anecdotes, largely of successful struggles, around which they structured their essays emotionally linked Lomonosov's life and the setting in which it took place to the products of his work and to his audience.

It was long assumed that Staehlin's "Traits and Anecdotes" was the biography that graced the 1784 edition of Lomonosov's collected works. Verevkin's authorship went unrecognized; indeed, to some he was thought to have been the translator for Staehlin's piece. This apparent error was rectified, with Verevkin receiving his due in recent decades.[36] It is clear, however, that Staehlin's 1783 biography, which began life as his undelivered 1765 eulogy to Lomonosov, appreciably shaped Verevkin's essay. Verevkin not only referred to Staehlin in his work, but also reprinted whole passages from Staehlin.[37] Verevkin did, however, provide certain "facts," not present in Staehlin, which have exerted considerable

---

[36]   The fact that at the beginning of Staehlin's study he indicated that it had been commissioned by Princess Dashkova for Lomonosov's collected works, to be issued in 1784, has naturally engendered confusion. Pekarskii, a skilled student of Lomonosov, thought that the essay was indeed Staehlin's work, and that it had been translated from the German by Verevkin (Pekarskii, *Istoriia Akademii nauk*, vol. 2, 259). M. P. Pogodin, who first published Staehlin's 1783 manuscript on Lomonosov in *Moskvitianin*, was convinced the author was in actuality D. S. -R. Damaskin, rector of the Slavo-Greco-Latin Academy and editor of the 1778 Moscow University edition of Lomonosov's works. See "Cherty i anekdoty," 1. On the basis of archival evidence pointing toward Verevkin's authorship, which he supplemented with textual comparisons between Staehlin's and Verevkin's works, D. S. Babkin in 1946 credibly laid claim on behalf of Verevkin. Babkin, "Biografii o Lomonosove," 12-27. Klaus Harer and Galina Smagina recently revisited this question. Harer holds that Verevkin served primarily as the (graceless) translator of Staehlin's essay: Harer, "Ia. Shtelin i M. V. Lomonosov," 188-90. Smagina (*Kniaginia i uchenyi*, 289-334), relying on Harer, equivocatingly postulates that Staehlin's biography in large part constitutes Vervkin's, albeit with substantive additions and modifications by Verevkin. Babkin's proofs authenticating Verevkin's primary responsibility for the composition, though too categorically drawn, remain authoritative.

[37]   The "derivative qualities" of the eighteenth-century biography is to be expected in the still-nascent genre. Korshin, "Intellectual Biography," 517.

sway over representations of Lomonosov's formidable precocity as an adolescent in the North. What Verevkin most depended on in Staehlin's memoir were the personal recollections that have been so instrumental in shaping the imagery of Lomonosov.[38] These "remembrances" provided some flesh to the skeletal chronological outlines of Lomonosov's official life.

Staehlin and Verevkin both make note of Lomonosov's plebian background as the son of a fisherman in the northern reaches of the country. The humbleness of his childhood (his father was actually a fairly well-to-do state peasant) would be greatly exaggerated in later historiography. In accompanying his father on fishing expeditions to the White Sea and along the Kola Peninsula, Lomonosov demonstrated to the authors an early example of the industriousness that became so associated with his name.[39] In his later papers, Lomonosov would vaguely allude to the sights seen on these travels in the North. This was often put forward as proof of Lomonosov's early fascination with nature, a first step on his road to the sciences.

According to Staehlin's account, repeated by Verevkin, Lomonosov, at the age of ten, "learned to read and write at the home of a neighboring deacon," who "knowing no Latin," was restricted to teaching young Lomonosov through "church books." But Lomonosov wanted to learn more, and was informed that for the "acquisition of more knowledge he would have to master Latin," which at that time he could learn in Moscow, Kiev, or Petersburg. It was also stated that "simple arithmetic [or calculations], he taught himself." This early mathematical referent, offered in varying forms by both Staehlin and Verevkin, became another source for considerable speculation about the origins of Lomonosov's scientific leanings.

All of this might seem to be merely indicative of a curious adolescent, but for someone emerging from such starkly "modest

---

38   Staehlin was also the author, or gatherer, of numerous anecdotes concerning Peter the Great, *Originalanekdoten von Peter dem Grossen* (Leipzig, 1785).

39   Shtelin, "Cherty i anekdoty," 1; Verevkin, "Zhizn' Lomonosova," III.

conditions," implicit is that curiosity to this degree has to have been an unequivocal exception from the norm. Staehlin provided no further information about Lomonosov's life prior to his journey to Moscow. Verevkin, though, writes that Lomonosov, in the home of a neighbor (Khristofor Dudin) "saw for the first time non-ecclesiastical books. They were the *Slavonic Grammar* [composed by Meletii Smotritskii early in the seventeenth century] and the *Arithmetic* [a mathematical textbook, with much attention given to navigational questions, authored by Leontii Magnitskii in 1703]." He would eventually inherit them after Dudin's death, and "from that time forward they never left him": he would "carry them throughout his life, rereading them over and over, indeed memorizing them," and he would later refer to them as his "gateways to learning."[40]

The *Slavonic Grammar* was, until the issuance of Lomonosov's *Russian Grammar* in 1757, the fundamental guide to language and rhetoric in Russia. Placed alongside Magnitskii's *Arithmetic*, which has long been proclaimed the principal published introduction to mathematics in eighteenth-century Russia,[41] and the *Psalter* of

---

40   Ibid., IV. Verevkin also stated that Lomonosov was for two years, from the age of thirteen, involved with religious dissenters (*Raskolniki*) who lived in his region—the far north was home to large numbers of *Raskolniki*. For a curious young person, and with dissenters so active in the area in which he grew up, this was hardly a surprising interest. Later, largely Soviet-era, historians, anxious to cast Lomonosov as a nearly lifelong materialist, tried to refute this claim of Verevkin's. For a studied effort to downplay Lomonosov's brief infatuation, see Morozov, *Put' k zrelosti*, 77-84.

41   Indeed, Magnitskii himself has long been revered in Russian scientific and cultural history; consequently, attempts to present his biography in a less hagiographic light have been rare. Recently, however, T. G. Kypriianova challenged his status as the sole author of the *Arithmetic* in her "Novye arkhivnye svedeniia po istorii sozdaniia 'Arifmetiki' L. Magnitskogo," in *Estestvennonauchnye predstavlenniia Drevnei Rusi*, ed. P. A. Simonov (Moscow, 1988), 279-82. She argues that the text was rather a collective undertaking, with Magnitskii playing a relatively minor part in the composition. Potentially more damaging to Magnitskii's reputation is research by W. F. Ryan. See his "Navigation and the Modernization of Petrine Russia: Teachers, Textbooks, and Terminology," in Bartlett and Hartley, eds., *Russia in the Age of the Enlightenment*, 75-105, which, convincingly, if still preliminarily, seeks to confirm that the unpublished mathematical

Simeon Polotskii, which Novikov argued that Lomonosov had read in his youth, it sketches before us an impressive beginning for young Lomonosov. That Lomonosov never mentioned the *Psalter* in his writings has already been established. Magnitskii's name is also not found among Lomonosov's many papers.[42] Lomonosov did utilize Smotritskii's *Grammar* in later life;[43] it is not possible, however, to confirm his knowledge of it in his youth. Whether or not Lomonosov was exposed to these materials in adolescence, and it seems unlikely that he was, is not of paramount importance. His reading of three texts vital to the evolution of seventeenth- and eighteenth-century Russian culture has been a fundamental ingredient of Lomonosov's biography for two hundred years. What were expected of figures of mythological or near mythological stature were early signs of greatness; these were the initial examples.

Substantial amounts of energy have been expended on trying to prove that during Lomonosov's lifetime the far north of Russia

---

manuscript of Henry Farquharson (a Scotsman who served forty years in Russia, commencing in 1699), and not the *Arithmetic*, was the primary such manual in eighteenth-century Russia. Additionally, Ryan submits that Farquharson, through his organizing and pedagogical activities first at the Moscow School of Mathematics and Navigation and then at the Naval Academy in St. Petersburg, was far more instrumental in establishing the basis for Russian scientific education than was Magntiskii. Perhaps proving the still totemic nature of Magnitskii's name, the aforementioned assertions by Ryan and Kypriianova are either ignored or inadequately dealt with in an otherwise finely detailed recent study of Magnitskii's *Arithmetic* by A. V. Lavrent'ev: *Liudi i veshchi. Pamiatniki russkoi istorii i kul'tury XVI-XVIII vv., ikh sozdateli i vladel'tsy* (Moscow, 1997), 78-80. Neither Ryan nor Kypriianova's revisionist positions are addressed in Okenfuss, *Rise and Fall of Latin Humanism*, 75-76, which continues to uphold the importance of the *Arithmetic* in Russian scientific history, as well as Magntiskii's pivotal role in the diffusion of mathematical knowledge.

42   Korovin thought that Lomonosov's familiarity with Magnitskii could be simply assumed, though he offered no proof for his supposition. Korovin, *Biblioteka Lomonosova*, 6-7, 65.

43   Lomonosov cited Smotritskii in such linguistic works as *Letter on the Rules of Russian Versification* (*Pis'mo o pravilakh Rossiiskogo stikhotvorstva*) and *Russian Grammar* (*Rossiiskaia grammatika*). See Lomonosov, *PSS*, vol. 7, 10-11, 412, 416, 597, 691.

was quite awash with all manner of freebooters, freethinkers, and foreigners, and therefore rich in nascent scientific ideas.[44] It would seem that Lomonosov, having grown up in an area of apparent intellectual ferment, had by a process of absorption assimilated elements of natural philosophy. This thesis is heavily dependent on the notions advanced by Novikov, Staehlin, and Verevkin. All that can be definitively asserted, however, is that given his acceptance into the Slavo-Greco-Latin Academy, he was to a degree literate.[45] Lomonosov began his four years (1731-35) at the Slavo-Greco-Latin Academy in Moscow after a "secret" departure that has since taken on the mantle of an epic flight from his father's home and village. Verevkin and Staehlin both outline this daring trip to a city where "he knew no one," with the sole goal of acquiring an education.[46] As a member of the poll-tax paying population, Lomonosov could not have gone without the permission of his village elders and without a passport. Although it was long ago verified that he had these,[47] this fact has barely made a ripple in the mythology, in which the legend of Lomonosov's secret flight still reigns.

The son of a peasant, Lomonosov, as is related by Verevkin and long since established as fact, had to lie about his origins—he claimed to be the son of a provincial nobleman—to gain admission

---

[44] On Lomonosov and the North, significant investigations include N. A. Go-lubtsov, ed., *Lomonosovskii sbornik* (Arkhangel'sk, 1911); Morozov, *Put' k zrelosti*, 1-99; and idem, *Rodina Lomonosova* (Arkhangel'sk, 1975), 331-83. See also Menshutkin, *Lomonosov: zhizneopisanie*, 1-9, and Shubinskii, *Lomonosov*, 13-62.

[45] There is also extant a fragmentary note, a legal contract of sorts, in Lomonosov's handwriting dating from 4 February 1726 (Lomonosov was fifteen). See N. A. Golubtsov, "Mikhail Vasil'evich Lomonosov," in Golubtsov, *Lomonosovskii sbornik*, 9; and Lomonosov, *PSS*, vol. 10, 479.

[46] Shtelin, "Cherty i anekdoty," 2; Verevkin, "Zhizn' Lomonosova," IV-V.

[47] Gur'ev, "Izvestie o Lomonosove," in *Puteshestviia akademika Ivana Lepekhina v 1772 godu*, part 4 (St. Petersburg, 1805), 302. Gur'ev was from the same district as Lomonosov, and apparently remembered him. In 1788, he provided fragmentary information on both the Lomonosov family and Lomonosov's legal exit from Kurostrov to the naturalist N. Ia. Ozeretskovskii, who then published it in the above volume.

to the Slavo-Greco-Latin Academy. When after three years or so Lomonosov's deception was discovered, Verevkin writes that Feofan Prokopovich (an influential clergyman and onetime close advisor to Peter the Great), who knew and appreciated Lomonosov's abilities, protected him from repercussions, saying, "Don't be afraid of anything, I will defend you."[48] As presented by Staehlin and Verevkin, Lomonosov in his zeal to advance his learning was willing to risk all. The obstacles in the form of social and legal impediments were overcome, and a powerful patron who recognized the young man's qualities and potential aided him in his quest. "In the monastery [The Slavo-Greco-Latin Academy was located at the Zaikonospasskii Monastery], Lomonosov studied with great diligence and achieved astonishing successes"—so much so, Staehlin and Verevkin comment, that he "successfully moved from the first class to the third class [the early classes focused on Latin, some Church Slavonic, and a smattering of history, both church and general] in six months."[49]

Verevkin, who had made investigations at the Slavo-Greco-Latin Academy, provided a relatively complete list of Lomonosov's courses and instructors at the Academy. In Lomonosov's "free time, instead of engaging in games like the other seminarians," he could be found in the Academy's library, where according to Verevkin, he would peruse "a few philosophical, physical and mathematical books" [no titles given]. Not finding enough in the library "to satisfy his hunger for learning," Lomonosov asked permission to go to the Kievan Academy, "for the study of philosophy, physics and mathematics."[50]

---

[48]   Verevkin, "Zhizn' Lomonosova," V-VI. Despite much research into the subject, it has not been proven that Lomonosov and Prokopovich ever met.

[49]   Ibid., VI; Shtelin, "Cherty i anekdoty," 3.

[50]   Verevkin, "Zhizn' Lomonosova," VI-VII; Shtelin, "Cherty i anekdoty," 3. It was long debated whether or not Lomonosov was actually ever in Kiev. Based on firm archival documentation, Galina Moiseeva has made a decisive case that he did visit the Kievan Academy during his time as a student at the Slavo-Greco-Latin Academy. G. N. Moiseeva, *Lomonosov i drevnesrusskaia literatura* (Leningrad, 1971), 75-7.

Unfortunately, Staehlin declares, he found nothing in Kiev but "arid fantasies instead of philosophy [or in Verevkin's similar account, hollow debates about Aristotle], and no possibility of achieving anything in physics or mathematics." Unable to find anything in the Kievan Academy's courses which attracted his interests, Lomonosov often visited its library where "for the lack of other books, [he] immersed himself in the chronicles and works of the church fathers," and then returned, early, to Moscow. It was not long after Lomonosov's arrival back in Moscow that the Slavo-Greco-Latin Academy, responding to a request from the Academy of Sciences for some students "who knew Latin, for the study of physics and mathematics with its professors," sent twelve of its number—including Lomonosov—to St. Petersburg. Verevkin and Staehlin tell us that in 1734 (actually 1735) Lomonosov, hearing of this possibility of advancing his studies at the Academy of Sciences, had persistently asked to be included in the group.

Verevkin's and Staehlin's biographies powerfully articulate the idea that Lomonosov's energy and innate abilities, which were, of course, already manifest during his early childhood, had flowered as far as was possible in the intellectually confining walls of the Slavo-Greco-Latin Academy. It is certain that Lomonosov gained competence in Latin, which was vital to a career in the sciences, in Moscow, along with mastery of Church Slavonic, as well as some exposure to ecclesiastical literature and an attenuated aristotelianism.[51] As for Lomonosov's fascination with

---

[51]   Little is known about the content of the natural philosophy taught at the Salvo-Greco-Latin Academy during Lomonosov's time in residence. V. P. Zubov, the best student of the topic, posited that "nothing favorable" had occurred there in terms of contemporaneous notions in "physics or experimental science": "Lomonosov i slaviano-greko-latinskaia akademiia," *Trudy instituta istorii estestvoznaniia i tekhniki* 1 (Moscow, 1954): 5-9, 46-52. See also Smirnov, *Istoriia Moskovskoi akademii*, 110-84, passim. My own investigations of Lomonosov's scientific papers and library do not reveal anything in the way of an intellectual debt in natural philosophy to the Slavo-Greco-Latin Academy. The Academy's first decade (it was directed from the late 1680s until 1694 by the Leichoudes brothers, Ioannikios and Sophronios) is the subject of a dissertation by Nikolaos A. Chrissidis ("Creating the New Educated Elite: Learning and Faith in Moscow's Slavo-

the oft-mentioned "physics and mathematics," its origin cannot be pinpointed. He did not receive any introduction to these subjects at either the Moscow or Kievan Academies.

Even Lomonosov's reading habits, central to both Staehlin and Verevkin's accounts, cannot, due to the 1737 fire that destroyed the requisite records at the Slavo-Greco-Latin Academy, be reconstructed. In writing their memoirs of Lomonosov, Staehlin and Verevkin, along with their generally correct skeletal outline of his life, were also signaling to their readers that Lomonosov's later great stature as a scientist grew from these early seeds. He surmounted with ease the impediments to his achieving success. The notion of intellectual obstacles overcome worked wonderfully in the mythology when combined with the economic and personal encumbrances in early life and at the Slavo-Greco-Latin Academy that Lomonosov described in his letters to Ivan Shuvalov.

The Lomonosov that Staehlin and Verevkin depicted as traveling to Moscow, Kiev, and St. Petersburg in the hope of acquiring more knowledge is squarely in line with the pattern that marked the written lives of many early natural philosophers. Lomonosov's most remarkable intellectual journeys took place in the Germanies. Before that, however, he was attached for a few months to the gymnasium that was part of the Academy of Sciences. Staehlin and Verevkin both erred in writing that Lomonosov was at the Academy for two years, with Staehlin writing that "with great success he studied physics and mathematics and also versification,"

Greco-Latin Academy, 1685-1694" [Ph.D. diss., Yale University, 2000]). Chrissidis makes the argument that while the Leichoudes brothers led the Academy its curriculum "acquainted the students both with the theoretical framework of natural philosophy, its vocabulary and terminology, as well as many of the latest advances in astronomy, albeit in a cursory manner, and very elementary conceptions of mathematics," and therefore the Academy "can be interpreted as the first attempt at institutional, formal education in science in Russia." Ibid., 267. What might have been true of the Academy's course offerings at the end of the seventeenth century, and Chrissidis's inferences are quite tentatively posed, seems clearly not to have been the case in the 1730s. Educational changes, or reforms, enacted during the reign of Peter the Great had both sharply altered and reduced the Slavo-Greco-Latin Academy's role in "educating the elite."

although none of his work has survived, "and he especially loved to immerse himself in mineralogy and experimental physics."[52] Lomonosov worked for a very short time in physics under Georg Krafft and in mathematics with Vasilii Adodurov, who was the first Russian adjunct at the Academy.[53] What Lomonosov picked up in the way of grounding in natural philosophy during these months is indeterminable and quite likely very little. No information of even the most indirect type has been discovered that would assist in answering the question.

In desperate need of people with skills in metallurgy and mining, the Academy decided to send three students (Lomonosov, Vinogradov, and Raiser) to train with the chemist Johann Henkel in Freiberg. First, though, they would need to master the basics of the sciences, and it was decided to have them study for a period with the Academy's good friend, the "famed philosopher and mathematician Christian Wolff."[54]

Although providing some exegeses on the literary work Lomonosov produced during his five years abroad, Staehlin, and through him Verevkin, is intent on placing Lomonosov's years of study in the more expansive framework of a young Russian wandering strange lands, often simply to where curiosity or circumstances led him. While he was journeying, Lomonosov's innate abilities and personal intellectual interests permitted him to master the array of scientific knowledge that would serve him so well later in life. The specifics of Lomonosov's education with Wolff were not vital, and were not dealt with; Lomonosov having been his pupil was the decisive factor. The minutiae of science were rarely probed in early memoirs of natural philosophers. That Lomonosov

---

[52]   Shtelin, "Cherty i anekdoty," 3. Staehlin arrived at the Academy before Lomonosov in 1735; his mistake is probably due to the fact that at the time he took no notice of the young Russian student's presence. Lomonosov may have also misinformed Staehlin at a later date. Except for an added reference to Lomonosov's studies in chemistry, Verevkin's report is essentially the same as Staehlin's. Verevkin, "Zhizn' Lomonosova," VII-VIII.

[53]   Chenakal, *Letopis'*, 31.

[54]   Shtelin, "Cherty i anekdoty," 3; Verevkin, "Zhizn' Lomonosova," VIII.

shared many of Wolff's views was not a point of debate among these eighteenth-century biographers.

Indeed, to a great extent Lomonosov's theoretical education came to an end after he left Marburg. Only much later, when Wolff's name became somewhat tarnished within Russia, did Lomonosov's scientific training in Marburg become suspect. After three years studying under Wolff, and "with his recommendation," Lomonosov moved on to Freiberg in order "to learn metallurgy and mining."[55] He was also mandated to observe the mining industry in Saxony. Verevkin and Staehlin state that Lomonosov returned to Marburg after a year—or approximately that long a period of time—to continue his study of the sciences. In truth he had, in the midst of serious quarrels with Henkel, and without the permission of the Academy of Sciences, left Freiberg in the spring of 1740, before his year had ended. Despite their conflicts, Henkel sent a positive evaluation of Lomonosov's progress with him to the Academy of Sciences, commenting that "Lomonosov has a very good knowledge of theoretical and practical chemistry, especially in metallurgy … and should be able to teach mechanics very well, for it is an area where, according to experts, he excels."[56]

With Lomonosov's departure from Henkel a year of adventurous travels throughout the German lands and Holland commenced as he, fearful that the Academy was angered by his actions, sought money and support for his return to Petersburg. Extreme material deprivations and a secret marriage that he kept hidden for some years defined for posterity this period in the young student's life.[57]

---

[55] Shtelin, "Cherty i anekdoty," 3; Verevkin, "Zhizn' Lomonosova," VIII-IX.

[56] Shtelin and Verevkin did not refer to this letter, which is reprinted in Lomonosov, PSS, vol. 10, 797. It is possible that wanting to ensure a continuing flow of Russian students, Henkel may not have wanted to upset the administration of the Academy of Sciences, and thereby perchance interrupt his financial support from the Academy, with a negative review of Lomonosov's abilities.

[57] Shtelin, "Cherty i anekdoty," 4-5, 8-10; Verevkin, "Zhizn' Lomonosova," IX-X, XII-XV.

A few months seem to have been spent "observing mines and smelting works" in Hessen, where he made the acquaintance of "the well-known mining and metallurgy specialist [Johan] Kramer" and "worked hard to learn these subjects."[58] From there Lomonosov traveled to Holland in order to gain the ear of Russian officials and ease his situation. En route, after imbibing too much at an inn along the way, he was impressed into the Prussian army.[59] Staehlin devotes considerable attention to this tale, and thrillingly narrates Lomonosov's escape. First Lomonosov made his way to Utrecht and later Amsterdam, where he found no one willing to assist him. His journeys finally came to an end when, after more time in Marburg, he embarked in the summer of 1741 upon a ship back to St. Petersburg. Staehlin inserts a vivid account, which is retold by Verevkin, of how on this voyage home Lomonosov had a dream in which he saw his father dead on an island in the White Sea. We are then informed that soon after Lomonosov's arrival his father's body was indeed found on the said island.[60]

These tales of adventure, danger, and prescience, which in no sense can be seen as ones experienced or shared by ordinary mortals, perfectly cast the young student as a hero in the making. Even when facing great pressures that would pull a lesser man from his path, Lomonosov would not, as Staehlin's and Verevkin's stories stress, be turned away from a vocation in the sciences. There is a nice amalgamation in these memoirs of pilgrimage and curiosity metaphors coexisting not in opposition to each other but rather supplementing each other in their mythic description of the youthful scientist's maturation. The descent of Staehlin's and Verevkin's biographies of Lomonosov from spiritual accounts of journeying is clear; they are, however, manifestly secular in their message and

---

[58]   Shtelin, "Cherty i anekdoty," 4-5; Verevkin, "Zhizn' Lomonosova," IX.

[59]   Shtelin, "Cherty i anekdoty," 5-7. Verevkin reproduced this story in his "Zhizn' Lomonosova" (see pp. X-XII). Lomonosov never wrote of his military "recruitment." He did state, however, that he had undergone very trying incidents on his travels.

[60]   Shtelin, "Cherty i anekdoty," 8-9; Verevkin, "Zhizn' Lomonosova," XIII.

represent the earliest written lives of a natural philosopher to appear in Russia. And in contrast to such contemporaneous evaluations as Novikov's, in these works Lomonosov's interests in chemistry, physics, and metallurgy, among other things, are interwoven with personal details in such a way that the reader could begin to perceive an association between his intellectual life and his biography.

Lomonosov also hinted at this association in a report sent by him, some months before his departure from Holland, to Johann Schumacher at the Academy of Sciences. Added to the locales he had passed through during his wanderings were Kassel, Leipzig, Frankfurt, The Hague and Rotterdam.[61] His desperate financial plight and the various forces, such as the enmity of Henkel toward him, which seemed to be conspiring to prevent him from pursuing an education, are amply covered. Lomonosov explained to an Academy at a loss as to his whereabouts the past few months, "of the dangers and want" he had lived through, "which are terrible even to remember." He wrote that he was then living "incognito" in Marburg, and even given all these hardships, "practicing his algebra, which he intended on using in chemistry and theoretical physics."[62]

The two biographies become somewhat perfunctory when addressing the last twenty-five years of Lomonosov's life at the Academy of Sciences, more a restating of titles and works than the story of a vibrant life. Staehlin noted that in 1746 (in fact it was 1745) Lomonosov was named a professor of chemistry and experimental physics (in truth, Lomonosov held the chemistry chair only). He "constructed a chemical laboratory [Russia's first in 1748] with the newest and best facilities, where he conducted many experiments and made many discoveries, which he outlined in papers read at

---

[61]   The letter is dated 5 November 1740. See Lomonosov, *PSS*, vol. 10, 421-31. Staehlin's and Verevkin's accounts differ slightly from Lomonosov's on the timing of his activities, though in general they complement each other. For a detailed chronology of Lomonosov's travails from the time he left Henkel to his shipping out from Lubeck, see Chenakal, *Letopis'*, 53-8.

[62]   Lomonosov, *PSS*, vol. 10, 430.

assemblies of the Academy."[63] Both authors went on to underline the amazing breadth of Lomonosov's published oeuvre, encompassing as it did panegyric speeches, poetry, the *Russian Grammar*, *Rhetoric*, dramatic works, studies in Russian history, and his manual on mining and metallurgy. Although the nature of Lomonosov's discoveries, experiments, and published scientific papers is not delineated, both biographers comment, not quite accurately, that it is possible to view them in detail in "his collected works and in the protocols of the academic Chancellery and Conference."

Staehlin and Verevkin remark that the great esteem Lomonosov was held in can be seen in the support proffered to him by luminaries in the Vorontsov and Shuvalov families, as well as, Staehlin writes, in the "respect given to him by many prominent European scientists and [scientific] societies, for example the Swedish and Bologna Academies, which elected him a member" (in 1760 and 1764 respectively, Lomonosov was elected an honorary member of said academies).[64] Strongly emphasizing the protean nature of Lomonosov's efforts, Staehlin insisted that these were "not anecdotes, but were all well-known deeds, and therefore, it would be a simple matter to collect more details about them."[65]

Lomonosov's endeavors in the development of mosaic art impress both authors,[66] who treat at considerable length an "anecdote" focusing on the seemingly fantastic scale of Lomonosov's work on such pieces as *Peter I* and the *The Battle of Poltava*. The science behind the mosaics is of little concern to them. That it was a product of prodigious energy and learning was self-evident to them, and, it is supposed, to their audience.

---

[63]   Shtelin, "Cherty i anekdoty," 10; Verevkin, "Zhizn' Lomonosova," XV.

[64]   Shtelin, "Cherty i anekdoty," 10-11; Verevkin, "Zhizn' Lomonosova," XV-XVI.

[65]   Shtelin, "Cherty i anekdoty," 11. In reviewing the scale of Lomonosov's many activities, Verevkin, who again used Staehlin's text almost verbatim, also argued that these were "not anecdotes, but were works [or accomplishments], and known everywhere." Verevkin, "Zhizn' Lomonosova," XV.

[66]   Shtelin, "Cherty i anekdoty," 11; Verevkin, "Zhizn' Lomonosova," XVI.

In dealing with Lomonosov's last years, to which Staehlin was a close witness, Staehlin and Verevkin were interested in advancing an image of Lomonosov's life and legacy which was unmatched in accomplishment at the time. Though Verevkin approached its veracity somewhat cautiously, Staehlin presented an anecdote that combines his high estimation of Lomonosov's place in Russian science and culture generally with Lomonosov's even higher self-representation. Staehlin apparently visited Lomonosov shortly before the latter's death. Worried about the sad state the sciences and learning would fall to in Russia without him, Lomonosov said to Staehlin:

> Friend [drug], I see that I will die and I look on death peacefully and indifferently. I regret only that I was unable to bring to completion everything I undertook for the benefit of my country, for the increase of learning and for the glory of the Academy, and now, at the end of my life, I realize that all my good works will die with me.[67]

This plaintive statement of Lomonosov's, possibly the most retold tale in the historiography, is central to myths of Lomonosov as the father of Russian science. Lomonosov's meeting his death bravely, concerned only about his legacy, the Academy of Sciences to which he devoted his adult life, and the country he loved, would become a leitmotif in every representation of him until the present day.

The Sentimentalist prose writer and poet Mikhail Murav'ev's "Contributions of Lomonosov to Learning," authored in the mid-1770s,[68] is the most scientifically informed of the eighteenth-

---

[67]   Shtelin, "Cherty i anekdoty," 12. In retelling Staehlin's tale, Verevkin substituted *priiatel'* for *drug* (Verevkin, "Zhizn' Lomonosova," XVII). *Drug* denotes a considerably closer relationship than *priiatel'*.

[68]   The dating of this piece, as with most of Murav'ev's writings, can only be approximated. (Murav'ev's archive at the Russian National Library in St. Petersburg contains more than forty volumes of manuscripts, the vast bulk of which were left in an incomplete state and remain unpublished.) Examinations of it (manuscript located at OR RNB, f. 499, ed. khr. 74, ll.

century biographies of Lomonosov and can be read as seamlessly capturing the scope of Lomonosov's reputation and image as a natural philosopher in the years immediately following the death of "Russia's first scientist." "Contributions of Lomonosov" has had a significant, though largely unexplored, role in structuring the mythology of Lomonosov in the history of Russian science.[69] Murav'ev was, according to Vadim Rak, a "connecting link" between such venerable littérateurs as Sumarokov and Nikolai Karamzin. Following Murav'ev's death Karamzin, who greatly respected him and owed his appointment as Imperial Historian to Murav'ev's intercession with Alexander I, edited a collection of Murav'ev's works. Murav'ev was also a celebrated member of the St. Petersburg cultural firmament to a degree that, for example, the soon-to-be-encountered Radishchev never was.[70]

1-9) and of the available biographical information on Murav'ev, and questions asked of scholars who have investigated Murav'ev's oeuvre, lead to the conclusion that he composed it in the mid-1770s. "Contributions of Lomonosov to Learning" ("Zaslugi Lomonosova v uchenosti") was first published in M. N. Murav'ev, *Opyty istorii, pis'men i nravoucheniia* (St. Petersburg, 1796), 132-39, then republished in *Opyty istorii, slovesnosti i nravoucheniia* (edited by Karamzin et al.), vol. 1 (Moscow, 1810), 180-90. It was also included in the collected works of Murav'ev issued in 1819-20 (three volumes) and reissued in 1847 (two volumes). Subsequent references will be to the 1796 edition.

[69]   It has been interpreted, usually quite cursorily, as simply part of a canon of worshipful biographical accounts. For two efforts at scrutinizing it, see Solov'ev and Ushakova, *Otrazhenie estestvennonauchnykh trudov Lomonosova v russkoi literature*, 16-17; and Zubov, *Istoriografiia*, 138-39.

[70]   Vadim Dmitrievich Rak, "Mikhail Nikitich Murav'ev (25 October 1757-29 July 1807)," in Levitt, *Early Modern Russian Writers*, 234. L. I. Kulakova was engaged for many years in collecting, analyzing, and publishing Murav'ev's poetry, and her "Poeziia M. N. Murav'eva," in *M. N. Murav'ev: stikhotvoreniia*, ed. L. I. Kulakova (Leningrad, 1967), 5-11, presents a superlative outline of Murav'ev's life. While Murav'ev has been the subject of considerable coverage, his "scientific" interests have received only meager attention. For more on Murav'ev, see also "Murav'ev (Mikhail Nikitich)—obshestvennyi deiatel' i pisatel' (1757-1807)," in *Entsiklopedicheskii slovar' (Brockhaus-Efron)*, vol. 20 (St. Petersburg, 1897), 189-90; I. Iu. Fomenko, "Istoricheskie vzgliady M. N. Murav'eva," *XVIII vek* 13 (1981): 167-84; *M. N. Murav'ev, Institutiones Rhetoricae. A Treatise of a Russian Sentimentalist*, ed. and with an introduction

Murav'ev had a long, albeit thinly documented, interest in the natural philosophy of his day. During his military service in St. Petersburg in the mid-1770s, he attended lectures in mechanics, mathematics (given by Euler), and physics given at the Academy of Sciences. Later he served as a tutor to the Grand Dukes Alexander and Constantine. He was particularly close to the future Tsar Alexander I, and in 1802 he was appointed his deputy minister of education. Active in the early "reform" period of Alexander's reign, he reinvigorated Moscow University, of which he was for some years the trustee.[71] Underscoring his attention to the sciences, he sponsored the establishment of such scholarly associations at Moscow University as the Moscow Society of Naturalists and the Society of Medical and Physical Sciences, both in 1805. What best attests to Murav'ev's scientific curiosity and knowledge, particularly in regards to the "first" Russian scientist, is, however, his writings on Lomonosov.

In addition to "Contributions of Lomonosov to Learning," Murav'ev published a lengthy panegyric to Lomonosov, "Eulogy to Mikhail Vasil'evich Lomonosov" ("Pokhval'noe slovo Mikhaile Vasil'evichu Lomonosovu," 1774).[72] Manifest in the essay is Murav'ev's enthusiasm for Peter I and his reforms.[73] He represented Lomonosov as embodying the spirit of Petrine transformations in Russian society. In his laudatory account, Murav'ev placed great

---

by Andrew Kahn (Oxford, 1995), XXII-XXXI; E. Petukhov, "Mikhail Nikitich Murav'ev: ocherk ego zhizni i deiatel'nosti," *Zhurnal Ministerstva narodnogo prosveshcheniia*, vol. 294, section 2 (August 1894): 265-96; and V. N. Toporov, *Iz Istorii russkoi literatury*. Vol. 2, *Russkaia literatura vtoroi poloviny XVIII veka: issledovaniia, materialy, publikatsii. M. N. Murav'ev: vvedenie v tvorcheskoe nasledie*, books 1-3 (Moscow, 2001-2007).

[71]   See A. Iu. Andreev, *Moskovskii universitet v obshchestvennoi i kul'turnoi zhizni Rossii nachala XIX veka* (Moscow, 2000).

[72]   "Pokhval'noe slovo Mikhaile Vasil'evichu Lomonosovu pisal leib-gvardii Izmailovskago polku kaptenarmus Mikhailo Murav'ev" (St. Petersburg, 1774).

[73]   An inquiry into Murav'ev's approach towards Peter the Great's impact on the course of Russian history is found in Fomenko, "Istoricheskie vzgliady Murav'eva," 181-83.

value on Lomonosov's labors in literature and the sciences and on his generally heroic example, which, given the nature of these types of pieces, it is not surprising he was unreservedly appreciative of, though he eschewed providing any details or analysis of Lomonosov's work as a naturalist.

Murav'ev also invoked Lomonosov's life in several shorter compositions, most interestingly in segments of "Three Letters" ("Tri pis'ma"). This was an early version of a genre favored by him, that of traveler's notes. In 1770-1771 Murav'ev, while visiting his father who was on government assignment in Arkhangel'sk, went on a "pilgrimage" to Lomonosov's birthplace near Kholmogory. "Three Letters" has a detailed description, almost spiritual in tone, of his impressions of the area that produced such a prodigy. Observing the remote village of "Kerostrov" [Lomonosov was born in Mishaninskaia on the island/district of Kurostrov], he recounted how "Having been absorbed in reading Lomonosov, [I found that] I am quite unable to gaze without passion and deference at the homeland of that brilliant mind."[74] The ethereal origins expected of a saint, secular or otherwise, are conveyed with appropriate rapture by Murav'ev.

Sentimentalism in Russian letters may have been partially motivated by a striving to overthrow the elaborate neo-classical

[74] *Sochineniia M. N. Murav'eva*, vol. 2 (St. Petersburg, 1847), 326. Irina Reyfman believes that Lomonosov's native village had become by the end of the eighteenth century a "place of worship." Reyfman, *Trediakovksy*, 96. Petr Chelishchev, a close friend of Radishchev, visited Kurostrov in 1791, not only leaving an informative chronicle of his travels but erecting a monument to Lomonosov in the area. See P. I. Chelishchev, *Puteshestvie po severu Rossii v 1791 godu. Dnevnik P. I. Chelishcheva* (St. Petersburg, 1886), 119-27. Like Murav'ev, Chelishchev asserted a connection between Peter the Great's actions and the rise of Lomonosov, a man so clearly a product of them. Thus was the deification of Lomonosov furthered by appealing to the more resonant memories of Peter the Great (on eighteenth-century identifications of Lomonosov with Peter the Great, see Zhivov, "Pervye russkie literaturnye biografii," 41). For data related to Chelishchev's travels in the North, see M. T. Beliavskii, "Petr Chelishchev i ego 'puteshestvie po severu Rossii'," *Vestnik Moskovskogo universiteta. Istoriko-filologicheskaia seriia*, no. 2 (1956): 40-47.

models formulated by Lomonosov and his imitators—Murav'ev indeed moved away from Lomonosov's example in literature,[75] though he never lost his unbounded respect for Lomonosov both as a potent cultural symbol and as a scientist of true ingenuity. The apotheosis of his interest in and ardent advocacy of Lomonosov's apparently prodigious legacy was "Contributions of Lomonosov to Learning." "Such an advantage is not bestowed on many minds," Murav'ev writes in introducing Lomonosov and his eclectic intellectual reach, "that combines an inclination for and abilities in fine arts with vast knowledge in the exact sciences: this superiority was possessed in the highest degree by our glorious compatriot, Lomonosov."[76] He stressed that Lomonosov took delight in pursuing all facets of learning; his horizons were not restricted to any one area of inquiry. There is nothing unexpected yet in Murav'ev's evaluation, which was typical of eighteenth-century—and nearly all later—memoirs focusing on the diverse character of Lomonosov's genius. While clearly cognizant of both the literary and scientific halves of Lomonosov's life, and sensitive to the impossibility of sharply dividing them, in "Contributions of Lomonosov to Learning" he concentrated on casting Lomonosov as the father of Russian science.

Drawing the reader's attention to chemistry, the area Murav'ev referred to as "the main occupation of Lomonosov," he praised Lomonosov's labors in mosaics, and the scienctific acuman needed to create the necessary colors—especially his work on *The Battle of Poltava*—and asserted that "besides his applications in this art, chemistry owes a debt of gratitude to him for his many

---

75   See Andrew Kahn's introduction to *M. N. Murav'ev, Institutiones Rhetoricae*, XXV-XXVI. N. D. Kochetkova, by contrast, is far less persuasive in not admitting a discernible evolution in Murav'ev's attitudes toward Lomonosov: "M. V. Lomonosova v otsenke russkikh pisatelei–sentimentalistov," in Kurilov, *Lomonosov i russkaia literatura*, 269-71.

76   Murav'ev, "Zaslugi Lomonosova v uchenosti," 132. The "exact sciences" that Murav'ev refers to in this first edition of "Zaslugi Lomonosova v uchenosti" is rendered as "physics and mathematics" in all later redactions.

important observations."[77] Impressed by both the content and style of Lomonosov's compositions in "chemistry," he then pronounced cogently on such treatises as "Metallurgiia" (*The First Principles of Metallurgy or Mining*); "O pol'ze khimii" (*Slovo o pol'ze khimii/ Oration on the Usefulness of Chemistry*, 1751); and "O proiskhozhdenii metallov" (*Slovo o rozhdenii metallov ot triaseniia zemli/A Discourse on the Birth of Metals from the Quaking of the Earth*, 1757),[78] which were all published in Lomonosov's lifetime and in Russian—as opposed to the far less accessible Latin of his more abstract, often unpublished expositions—and were, as already illustrated, among the best known of his papers.

Murav'ev, like Radishchev later, found the geological processes described by Lomonosov, which led to the formation of islands, mountains, and "in unattainable depths to treasures of gold and silver," to be exceptionally noteworthy. This was also, in varying ways, a fairly pragmatic research product with greater resonance in Russia than the corpuscular dissertations that were, by means of the Academy's Latin scientific journal(s), received and criticized in Western Europe. Murav'ev held up Lomonosov, who in "his zeal for the good and glory of his native land proposed useful projects,"[79] as the foremost exponent in Russia of practical scientific progress. He did not neglect Lomonosov's more theoretical excursions, but in a wider social climate ill-disposed to the reception of unorthodox ideas, sketching out areas where Lomonosov's work had, or might offer in the future, quantifiable results was perhaps a sound way for him to extol Lomonosov's overall scientific legacy.

Holding out the potential of immediate benefits, particularly to a nation with problematic outlets to the oceans, were advances in geography, geodesy, and related sciences. Lomonosov's attempts to plot a northern sea route to the "East," or India, which he turned to several times in the 1750s and early 1760s,[80] was the

---

[77]   Ibid., 136-37.

[78]   Ibid., 137.

[79]   Ibid., 135.

[80]   *A Discourse on Greater Accuracy of the Sea Route (Rassuzhdenie o bol'shei*

topic of abundant coverage by later generations of scholars who often argued, rather creatively, that Lomonosov anticipated later navigational and cartographic findings. Although Lomonosov's search for a northern sea route looked as though it came to naught, Murav'ev was not willing to concede that anything that Lomonosov spent prolonged time on could be bereft of any positive result. He contended that Lomonosov's conjectures were ahead of their time, arguing, for example, that Lomonosov's projects seemed to "situate America closer to Russian dominions than depicted on contemporaneous maps."[81] Murav'ev also maintained that later navigational investigations confirmed Lomonosov's prescience. His decision to award this discovery of sorts to Lomonosov signals the early roots of what became a constant theme in Russian treatments of Lomonosov's science: the contestation for priority over discoveries. To many adherents of Russian precedence in an array of scientific fields, the credit too often seemed to go to non-Russian scientists at the expense of their rightful recipient, Lomonosov.

Expanding the parameters of Lomonosov's scientific legend, Murav'ev advanced two theses further articulating the idea of Lomonosov as an independent thinker of marked inventiveness, which were regularly revisited by subsequent students of

---

tochnosti morskogo puti), read by Lomonosov at a public assembly of the Academy of Sciences on 8 May 1759, was the best known of his navigational/geographic papers. See Lomonosov, *PSS*, vol. 4, 123-86, 740-59. Published in 1759 in both Russian and Latin, it was included in the 1778 and 1784-87 editions of Lomonosov's collected works. See *Svodnyi katalog*, vol. 2, 165-66, 173. In addition to this published navigational paper, Murav'ev's discussion indicates some familiarity with Lomonosov's *A Brief Description of Different Voyages in the Northern Seas and Evidence of a Possible Route through the Siberian Ocean to East India* (*Kratkoe opisanie raznykh puteshestvii po severnym moriam i pokazanie vozmozhnogo prokhodu Sibirskim okeanom v Vostochnuiu Indiiu*). First published in 1847 by the geographer A. P. Sokolov, it was reissued by him in 1854 along with several other documents related to the so-called Chichagov expedition. A. P. Sokolov, ed. *Proekt Lomonosova i ekspeditsiia Chichagova* (St. Petersburg, 1854). In the months before his death, Lomonosov had been deeply involved in the preparations for the Chichagov voyage.

81   Murav'ev, "Zaslugi Lomonosova v uchenosti," 135.

Lomonosov's science. First, he unhesitatingly accepted Lomonosov's equivalence to Benjamin Franklin, bitingly disputed later by Radishchev—indeed, his arguments are insistent in pressing for their shared provenance over electrical experiments and discoveries; second, he determined that in certain of his hypotheses falling under the heading of physics, Lomonosov was as original as Newton.

Paying heed to the fame accrued by Lomonosov's late colleague, Richmann, Murav'ev associated Lomonosov with him, and hence with the highly publicized electrical experiments that "occupied the attention of thinking people from one end of Europe to the other"[82] that led to Richmann's death. Lomonosov did work closely with Richmann, but Murav'ev also knew how best to augment Lomonosov's stature. Even more sensitive to the desirability of equating Lomonosov with Franklin, Murav'ev, in a rebuff to would-be skeptics, hoped to persuade his readers that:

> At the same time, as the celebrated American, Benjamin Franklin ... through many curious observations paved for himself the road towards ingenuously solving the problem, in what way the circulation of the invisible force, which is dispersed everywhere, attracts or repels bodies, Lomonosov by the power of his own reasoning had reached the same conclusions and shared with him the glory of the invention.[83]

Murav'ev did not question the significance of Franklin's exertions, but he viewed Lomonosov's efforts, which were conducted almost simultaneously with Franklin's, as completely autonomous and equal in result.

Murav'ev then posited the analogous example of the concurrent work at the beginning of the century conducted by Leibniz and Newton on developing the calculus. Though not made explicit, the implication is clear: Newton's gigantic, and ever growing, stature had threatened to block out recognition of Leibniz's seminal role.[84]

---

[82]   Ibid., 133.

[83]   Ibid.

[84]   Mindful of the bitter dispute over credit for inventing the calculus that

Alluding to Lomonosov's paper(s) on electricity, he commented that, "Like Leibniz, Lomonosov sensitively affirmed for himself his share of the discovery and proved in his own particular writing, that he does not owe his thoughts to Franklin."[85] Murav'ev judged this as having profound consequences for both Lomonosov's legacy and his country, for "this ambition pertaining to his fatherland convinced other countries, which had an earlier enlightenment, that Russia had matured into a rival with them in the dissemination of scientific knowledge.

Murav'ev held that Lomonosov accomplished his more substantive work in physics, rather than in chemistry. It was after all physics, not chemistry, that had mainly benefited from the eighteenth-century advances in mathematics, thanks to the achievements of Newton, Leibniz, Euler, et al., and where the most exciting advances took place. Physics, Murav'ev notes, "repeatedly drew the attention of Lomonosov, and in certain spheres, he brought to bear the assistance of mathematical calculations [or more precisely, calculus]."[86] Lomonosov's treatises falling under the rubrics of either physics or chemistry—the vast majority of his work—do not, however, reveal his command of mathematics, but rather his failure, or inability, to utilize it in his work beyond a most rudimentary level.

At the end of the nineteenth century the emergence within scientific disciplines of trained individuals devoted to studying the history of their profession, a development which signaled a certain maturation of the genre of scientific biography, led to the minutiae of Lomonosov's science becoming the focus of sustained scrutiny. With this more rigorous examination, the discernible gaps in Lomonosov's learning became issues that required greater sophistica-

---

broke out at the beginning of the eighteenth century between Newton and Leibniz which, combined with Leibniz's generalized opposition to what would become the dominant scientific worldview, Newtonianism, had the effect of seriously eroding Leibniz's scientific reputation, Murav'ev strove to prevent a similar fate from being visited on Lomonosov.

85   Murav'ev, "Zaslugi Lomonosova v uchenosti," 134.

86   Ibid., 138.

tion. Until this time it was the exemplary life of the scientist, rather than the intricacies of experimentation and discovery, that was seen as most relevant to the writing of his life.

In any case, while Murav'ev likely did not have the scientific rigor to engage with the more exacting developments in the sciences, which in the late eighteenth century meant the application of advanced mathematics to physical phenomena, he did understand that mathematics had become a requisite part of the scientific armor of any "modern" natural philosopher, and he argued for its importance to Lomonosov's science. Whether Murav'ev's assertion was the result of a misreading of Lomonosov's scientific skills, incomprehension of mathematical analysis, or instead because he was writing within an already powerful mythical tradition that did not permit deviation, is debatable. What is indisputable is that the notion of Lomonosov as a natural philosopher with a sure grasp of mathematics became a significant, though not completely uncontested, element in representations of Lomonosov over the following two centuries.

Placing Lomonosov near Newton in the pantheon of renowned scientists was the most audacious aspect of Murav'ev's account. He esteemed Lomonosov's imaginative efforts to sketch out a mechanical theory of light and color highly, and endeavored to convince others that:

> Bringing to all of the sciences a bold spirit of experimentation, he had the courage to differ with Newton about the origin of light and the attributes of colors. He inferred that the appearance of light and heat are determined by the rotary motion and quick revolution of the intangible parts of bodies, which he termed ether, and proposed that colors are produced from the interaction of different parts of the ether with parts of mercury, sulphur and salt.[87]

He is conversant with Lomonosov's optical treatise, *Oration on the Origins of Light, Representing a New Theory of Colors* (Slovo o pro-

---

[87]   Ibid.

iskhozhdenii sveta, novuiu teoriiu o tsvetakh predstavliaiushchee),[88] in which the scientist sharply distinguished his conclusions, which, simply put, can be classified as adhering to wave theories, from what he perceived to be the strictly corpuscular bases of Newton's emission theory of light. Although Lomonosov's dissertation was not based on the "bold" experimentation lauded by Murav'ev's rhetoric, in fact a continuing tension in later studies of Lomonosov concerns the degree of experimentation that he engaged in to support his speculations in physics and chemistry; in this instance, he did at the very least exhibit originality. In any case, Murav'ev seemed less interested in the intrinsic value of Lomonosov's research than in the fact that he offered hypotheses which challenged those of Newton.

His assessment might at first glance lead to the unfounded assumption that in Russia Newtonianism, at least in the field of optics, and in the opinion of as learned an observer as Murav'ev, had yet to triumph over competing views, like those expressed by Lomonosov.[89] But whatever the validity of Lomonosov's theorizing, Murav'ev does not suggest an exact equivalency between Newton and Lomonosov, and despite the incontrovertible evidence that he gleaned from Lomonosov's writings he did not depict him as an obstinate foe of Newton's ideas; indeed in another article, "Eloquence" ("Krasnorechie"), Murav'ev, alluding to Lomonosov's work in investigating light, explicitly enlisted him, quite erroneously, as

---

[88]   Lomonosov, *PSS*, vol. 3, 315-44, 550-55.The Russian version was first published in 1758, followed the next year by a Latin translation. For its various eighteenth-century reprintings, see *Svodnyi katalog*, vol. 2, 163-66.

[89]   While Newton's name had become nearly inviolate in European scientific circles by the end of the eighteenth century, there was nothing inevitable about the progression of Newtonian influences, and their advancement varied greatly over time and place. For the tensions between Newtonian, or emission, and—for want of a better term—Eulerian, or wave, theories, and the variegated fate of these theories in Russia, see Boss, *Newton and Russia*, 156-59, 185-98. R. W. Home's "Leonhard Euler's 'Anti-Newtonian' Theory of Light," *Annals of Science* 45, no. 5 (September 1988): 521-33, though restricted in scope to criticisms of Newton made by Euler during his Berlin sojourn (specifically in 1744-46), should, by virtue of Euler's great sway over Russian scientific thought, be consulted.

a "follower of Newton's."[90] He evidently wanted Lomonosov to be seen as worthy of being associated in history with revered scientists such as Newton.[91]

Even so, anticipating in a sense Radishchev's perception that Lomonosov's image was his most valuable gift to his country, he concluded "Contributions of Lomonosov to Learning" by declaring that "Lomonosov belongs to that small number of inventive minds, and by his own example affirms the truth, that Russians are endowed with great intellectual abilities."[92]

The heroic self-image Lomonosov expressed in his letters to Shuvalov exhibit, as indicated earlier, close affinities with the autobiographical musings of other early modern scientists. That they were the first such pieces composed by a Russian gives them a foundational aura not only in the historiography devoted to Lomonosov but also in Russian science. The hagiographic early biographies of Lomonosov combined with the mythogenic tendencies of post-Petrine Russia to configure him as a Russian Newton, Galileo, Copernicus, or Franklin. It was an image that Aleksandr Radishchev, a writer viewed reverently by many later

---

[90]   *Sochineniia Murav'eva*, vol. 2, 246. In the same passage Murav'ev continued to maintain that, in laying bare the laws of electricity, Lomonosov was a "rival of Franklin's."

[91]   Writing at about the same time as Murav'ev, the poet Semën Bobrov—an acquaintance of Radishchev and admirer of Lomonosov's poetry and science—in an effort to elevate Lomonosov's prestige substituted Lomonosov's name for Newton's in a passage of his poem *Tavrida* (1798), which alluded to the discovery of the sun's spectrum (and therefore to Newton's *Opticks*). Bobrov's composition was modeled in part on James Thomson's *The Seasons*. This reference is taken from Iu. D. Levin, *Vospriiatie angliiskoi literatury v Rossii* (Leningrad, 1990), 199.

[92]   Murav'ev, "Zaslugi Lomonosova v uchenosti," 139. The original passage reads "Lomonosov prinadlezhit k malomu chislu dukhov sotvoritelei i dovol'no odnovo ego, chtob osnovat' preimushchestvo velikoi sposobnosti Rossiiskago dukha." Murav'ev's rather awkward Church Slavonic and Russian construction was modernized by Karamzin as "Lomonosov prinadlezhit k malomu chislu umov izobretatel'nykh i odnim primerom svoim utverzhdaet istinu, chto rossiiane odareny velikimi sposobnostiami razuma." The latter rendition appears in all subsequent reprintings of "Zaslugi Lomonosova v uchenosti."

Russian and Soviet scholars as Russia's first "intelligent" and even "revolutionary," struggled against. Radishchev's stature in Russian culture, which was decidedly modest in his own lifetime, eventually rose to a level second only to those of Pushkin and Lomonosov. That he took issue with the canonization of Lomonosov is alone worth noting; that his evaluation of Lomonosov's achievements is so at variance with what came before makes it unique.

Notwithstanding the copious literature on Radishchev, it has proven very difficult to decisively delineate the origins of his intellectual biography.[93] Radishchev's interests in the sciences of his day can be demonstrated: he was familiar with Lomonosov's scientific papers (at least the ones published in the 1784-87 edition of Lomonosov's collected works) and, as his work on him indicates, he eloquently assessed their merit, but the precise lineage of his inquisitiveness is unclear.[94] Although he was skeptical of aspects of the new dominant Newtonian worldview, and a believer in phlogiston—not a remarkable position for the time—his evaluation of Lomonosov's legacy in the sciences remains intuitive.

Radishchev's "Discourse on Lomonosov," the last chapter of his famous *Journey from Petersburg to Moscow* (*Puteshestvie iz Peterburga v Moskvu*, 1790),[95] serves not only as a substantive early

---

[93]   G. P. Makogonenko's study adduces a spectacularly protean intellectual development for Radishchev, but it is far too speculative to be accepted uncritically. See his *Radishchev i ego vremia* (Moscow, 1956), 3-121. For a corrective, see Allen McConnell, *A Russian Philosophe: Alexander Radishchev* (The Hague: M. Nijhoff, 1964), 1-40.

[94]   P. A. Radishchev, "A. N. Radishchev," *Russkii vestnik* 18, book 1 (December 1858): 58, 399-401; P. M. Luk'ianov, "A. N. Radishchev i khimiia, *Trudy Instituta istorii estestvoznaniia i tekhniki* 2 (1954): 158-67; Raskin, *Khimicheskaia laboratoriia*, 211, 279; Zubov, *Istoriografiia*, 91-103; Boss, *Newton and Russia*, 227-28, 236. In one of his essays Radishchev did indicate some awareness of the experiments conducted by Joseph Priestley. See his "O cheloveke, o ego smertnosti i bessmertii," in A. N. Radishchev, *Polnoe sobranie sochinenii*, vol. 2 (Moscow, 1941), 78-79, 81, 92.

[95]   The *Journey from Petersburg to Moscow* is one of the best known literary works in Russia. Although both the sources on which Radishchev based his work and its political and social messages have been subject to considerable debate, the *Journey*, written to a degree in the manner

evaluation of Lomonosov's science but as an attempt to re-channel the growing idea of Lomonosov as the father of Russian science, on a par with the most celebrated experimenters of his day, into a less exalted framework. To Radishchev, the mythology that had developed around Lomonosov threatened to become starkly disproportionate to his actual accomplishments.

Radishchev's "Discourse on Lomonosov," which he worked on intermittently between 1780 and 1788, though included in his *Journey*, was conceived as an independent work, and stands on its own. Students of Lomonosov have referred to Radishchev's essay, or at the very least juxtaposed his name with Lomonosov's, in what would appear to be every major study of the scientist. Rare, however, are those who give more than a cursory glance at Radishchev's critique of Lomonosov's science.[96] Due to the perceived negative

---

of Laurence Sterne's *A Sentimental Journey*, can be interpreted as a plea for immediate internal reforms and an attack on serfdom. Reacting with fear to events in revolutionary France, Catherine II was outraged by the publication of the book—particularly bothersome to her was the fact that the censors had granted permission for an initial rendering of the *Journey* prior to its publication—and the Russian government initially sentenced Radishchev to death. His punishment was later commuted to internal exile. For the most recent "definitive" version of the "Discourse on Lomonosov," see *Puteshestvie iz Peterburga v Moskvu. Vol'nost'*, ed. V. A. Zapadov (Leningrad, 1992), 115-23, 463-72. For the complex history of Radishchev's compilation and revisions of his *Journey*, which touches only sporadically on the "Discourse," see V. A. Zapadov "Istoriia sozdaniia 'Puteshestvie iz Peterburga v Moskvu i 'Vol'nost'," in ibid., 475-560.

[96] The only "recent" Russian studies that exclusively investigate the "Discourse on Lomonosov" are tendentious analyses focusing mainly on Radishchev's opinions of Lomonosov's literary and linguistic efforts, which see his essay as nearly unequivocal in its admiration of Lomonosov. See L. I. Kulakova, "A. N. Radishchev o M. V. Lomonosove," in *Literaturnoe tvorchestvo M. V. Lomonosova: issledovaniia i materialy*, ed. P. N. Berkov (Moscow-Leningrad, 1962), 219-36; V. I. Moriakov, "A. N. Radishchev o M. V. Lomonosove," *Vestnik Moskovskogo universiteta*, series 8, history, no. 4 (July-August 1986): 34-43; Il'ia Z. Serman, "<<Slovo o Lomonosove>> i ego mesto v <<Puteshestvii iz Peterburga v Moskvu>>," in *Problemy izucheniia russkoi literatury XVIII veka: Mezhvuzovskii sbornik nauchnykh trudov*, ed. E. I. Annenkova and O. M. Buranok (Samara, 2001): 222-32; and A. G. Tatarintsev, "'Slovo o Lomonosove' A. N. Radishcheva. (K probleme tvorcheskoi istorii 'Puteshestviia')," in *Voprosy russkoi i zarubezhnoi literatury*

nature of Radishchev's views, they are usually dismissed as either a result of Radishchev's ignorance of the extent of Lomonosov's scientific output or an inexplicable flaw in the "Discourse." Galina Pavlova echoed the predominant opinion of Russian, Soviet, and Western scholars who have looked at Lomonosov and Radishchev in asserting that because most if not all of Lomonosov's papers were unpublished or otherwise forgotten, Radishchev could not properly assess hiss theoretical researches.[97] The "Discourse on Lomonosov" then is one of the staples in the Lomonosov canon, but one that due to its apparent ambiguity in appreciating Lomonosov's talents as a natural philosopher has been approached wholly inadequately.

In the "Discourse," Radishchev described Lomonosov's early years and education in Moscow and Western Europe in the common fashion of portraying it as a pilgrimage, with many valuable diversions along the way, to knowledge. Armed with the basics of Lomonosov's biography, and with a strong acquaintance with the works of Staehlin, Verevkin, and Novikov, Radishchev treated

---

(Perm', 1974), 17-36. P. M. Luk'ianov attempted the most detailed work on Radishchev as a natural philosopher; beyond some perfunctory quotations from the "Discourse," however; he evaded an explication of Radishchev's appraisal of Lomonosov's science. See his "Radishchev i khimiia," 165. Andrew Kahn's Bakhtinian analysis of the *Journey* briefly touches on the "Discourse." Although Kahn's attempt to provide a more multivalent interpretation of the text and to insert a presumed dialogic relationship between the narrator—Radishchev—and Lomonosov is often compelling, his conclusion that Radishchev viewed Lomonosov as a "secular deity," even if only as a littérateur, is exaggerated. See Andrew Kahn, "Self and Sensibility in Radishchev's *Puteshestvie iz Peterburga v Moskvu*: Dialogism and the Moral Spectator," *Oxford Slavonic Papers* 30 (1997): 65.

[97]   Pavlova, *Lomonosov v vospominaniiakh*, 10-11; and idem, "Lomonosov v kha-rakteristikakh," 68-69. See also Babkin, "Biografii o Lomonosove," 46-47; Kulakova, "Radishchev o Lomonosove," 235. Alexander Vucinich contended that, "What Radishchev thought of Lomonosov as a scientist is not today of any great significance, and we must remember that he did not have access to Lomonosov's scientific papers." Vucinich distinctly qualified this, quite accurately I believe, by also stating that in general terms Radishchev's "opinion at the time helped to clarify Lomonosov's true place in the history of Russian science." Vucinich, *Science in Russian Culture: A History to 1860*, 115.

Lomonosov's presumably surprising emergence from distant
Kholmogory and the intellectual deprivations he overcame with
wonderment. He saw Lomonosov's possession of a preternatural
curiosity as his most striking trait, a curiosity that "strives after the
knowledge of things…. It roars, seethes and groans, and breaking
its bonds in an instant, flies headlong (nothing can stop it) toward
its goal." In the face of this, "Everything is forgotten, there is only
one object in mind; by it we breathe and live." Beyond all other
considerations, his aspiration was the "knowledge of things."[98] This
yearning of Lomonosov's could not be satisfied in Russia, so he
traveled to Marburg, where:

> He became a student of the famous Wolff. Rejecting the
> rules of scholasticism, or rather the errors taught him in the
> monastic schools, he laid down firm and clear steps that led
> up to the temple of philosophy. Logic taught him to reason;
> mathematics taught him to draw sound conclusions and to be
> convinced only by firsthand evidence; metaphysics instructed
> him in conjectural truths which often lead to error; physics
> and chemistry, to which he devoted himself eagerly, perhaps
> because of their pleasant stimulus to the imagination, led
> him to the altar of nature and disclosed its mysteries to him;
> metallurgy and mineralogy, as corollaries of the preceding
> subjects, attracted his attention, and Lomonosov tried eagerly
> to learn the laws which governed these sciences.[99]

Throughout the *Journey* Radishchev rails against the abuses
long endured by the Russian peasantry. A true philosophe, admiring of, if not always completely conversant with, the latest scientific developments in the West—as his rejection of Newtonianism
makes clear—Radishchev perceived that the arrival of modern natural philosophy in Russia could foreshadow generalized cultural
reforms. The image of this son of the peasantry traveling through

---

[98]   Aleksandr Nikolaevich Radishchev, *A Journey from St. Petersburg to Moscow*,
        ed. Roderick Page Thaler, trans. Leo Weiner (Cambridge, MA: Harvard
        University Press, 1958), 224.

[99]   Ibid., 226-27.

exotic places in an attempt to master something even more exotic for an eighteenth-century Russian, the sciences, was greatly valued by Radishchev. Wolff's reputation had not dimmed in Radishchev's eyes, and his association with the young Lomonosov was noteworthy. Radishchev was unusual among those who wrote about Lomonosov both then and since in eschewing the anecdotes that Staehlin had implanted in the historiography.[100] He did not reiterate all of the well-known heroic tales attached to Lomonosov's life, but instead was animated by the myth of Lomonosov itself, which he attempted to dissect, and asked what Lomonosov's bequest to the sciences was. In doing so, Radishchev insisted that "we want to do justice to the great man but not to imagine that he was God the Creator of all; let us not set him up as an idol to be worshiped by society nor contribute to the establishment of any prejudice or false conclusion."[101]

In one of the longest passages in the "Discourse" Radishchev depicts both the horrors and the usefulness of mining.[102] Utilizing knowledge gained from Lomonosov's "On the Strata of the Earth" ("O sloiakh zemnykh"),[103] which was one of the supplements to *The First Principles of Metallurgy or Mining*, he provides a vivid geological description of an excursion through the subterranean world. It was for these more "practical" labors in mining, metallurgy, and geology that Radishchev thought Lomonosov should be memorialized. While this emphasis of Radishchev's may in fact depend on questions of accessibility, both to the actual texts of Lomonosov's papers in chemistry and physics and to Radishchev's understanding of their

---

[100]   Kulakova considered Radishchev's exclusion of Staehlin's anecdotes one of the aspects of the "Discourse" that made it the premier "contemporary" memoir of Lomonosov. This seems more motivated by her animus toward Staehlin's influence in Lomonosov studies than a well-thought analysis of Staehlin's formative role in creating the mythology of Lomonosov. Kulakova, "Radishchev o Lomonosove," 225-28.

[101]   Radishchev, *Journey*, 235.

[102]   Ibid., 227-29.

[103]   "O sloiakh zemnykh," issued in 1763, was also included in *Polnoe sobranie sochinenii Lomonosova*, vol. 4, 1785, 168-294.

content, he did not neglect Lomonosov's more abstract work; he simply assessed it as less than pioneering in impact.

Arguing against too ready an equivalence between Lomonosov and selected august scientists, Radishchev reasoned:

> Nor will we place him on the same level with Margraf or Rudiger [Andreas Sigismund Margraff and Johann Andreas Rudiger were prominent eighteenth-century German chemists], merely because he worked at chemistry. Though this science fascinated him, though he spent many days of his life in the investigation of the truths of nature, his course was that of a follower. He walked on trails previously opened up, and in the endless riches of nature he did not find the smallest blade of grass that better eyes than his had not seen, nor did he find any more primitive sources of matter than his predecessors had discovered.[104]

Margraf and Rudiger had prestigious pan-European reputations, something that Lomonosov never enjoyed. Radishchev judged Lomonosov as less than original in the substance of his theoretical conclusions, a jarring evaluation in the eyes of many later scholars of early Russian science, who credit Lomonosov with precedence over a multitude of discoveries. And in a comparison that brought forth much consternation from later writers, Radishchev inquired of his readers:

> Shall we place him near the one who was honored with the most flattering inscription any man could see beneath his portrait? It is an inscription not etched by flattery, but by truth attacking tyranny: 'He has snatched the lightning from heaven and the scepter from the hands of tyrants [Turgot's epigram (1778) on Benjamin Franklin].' Shall we place Lomonosov near him because, having investigated electricity, he knew how to ward off the thunderbolt, but in this science Franklin is the architect, Lomonosov an artisan. But if Lomonosov did not achieve greatness in the investigation of nature, he described its marvelous workings in a pure and understandable style. And although his works on natural science do not show

---

[104]   Radishchev, *Journey*, 236.

him to be a master scientist, we find him to be a master of expression and always an example worthy of emulation.[105]

Although it was not a crucial component of his evaluation, Radishchev was rankled by Lomonosov's apparent flattery and seeking after the patronage of high-ranking personages. In this regard, and in his stature as a natural philosopher, Radishchev consigned Lomonosov to a lesser place among the immortals than he did Benjamin Franklin. As has been remarked earlier, Lomonosov's research in electricity and the subsequent publication of his findings had perhaps the widest contemporary reception of all of his works. Whether priority in these electrical experiments belongs to Franklin or Lomonosov is less relevant than is the fact that Radishchev's contention is utterly at odds with what became almost an axiom in the literature: that Lomonosov either anticipated Franklin's results or came to them independentaly of Franklin, though they were equal in import.[106]

---

[105]   Ibid.

[106]   The source of this claim can be traced to Lomonosov himself. Despite his acquaintance with Franklin's electrical experiments (Lomonosov obliquely cited Franklin's *Experiments and Observations on Electricity, Made at Philadelphia in America*, 1751), he disavowed any notion that he was indebted to his work, writing that "in my theory about the cause of electric power in the air I owe nothing to him, as is apparent from the following...." Lomonosov contended that "I only saw Franklin's writings when I had already prepared my speech." And after listing what he perceived to be gaps in Franklin's research, which he argued resulted largely from Franklin's observations having been made in a far different climate than St. Petersburg's, he concluded, "there are many phenomena related to thunder [and related atmospheric changes], which in Franklin's work there are no traces of." See Lomonosov's 1753 paper, *Explanations, Required for a Word about Electrical Air Phenomena* (*Iz'iasneniia, nadlezhashchie k slovu o elektricheskikh vozdushnykh iavleniiakh*), in Lomonosov, *PSS*, vol. 3, 103 (101-133 for entire treatise). This paper was a supplement to his *Discourse about Air Phenomena, Caused by Electricity* (*Slovo o iavleniiakh vozdushnykh, ot elektricheskoi sily proiskhodiashchikh*). See ibid., 15-99, 512-22. Also composed and published in 1753, it was reprinted in the 1784-87 edition of Lomonosov's collected works consulted by Radishchev. For the several eighteenth-century incarnations of these two dissertations, see *Svodnyi katalog*, vol. 2, 163-66, 175-76.

Radishchev deemed Lomonosov's literary and linguistic accomplishments more significant than his scientific output, which he saw as more innovative in style than in substance. He perhaps too readily compared Lomonosov's science with that of natural philosophers working in more scientifically hospitable climes. It is pertinent to observe that while Lomonosov knew of Franklin's work, Franklin also knew of Lomonosov's investigations. In early 1765, at the behest of Ezra Stiles—then a minister in Rhode Island and a naturalist, and later president of Yale University—Franklin agreed to dispatch correspondence requesting information on Russian meteorological conditions to Lomonosov from London. Stiles also inquired what Franklin knew, or could find out, of Lomonosov's efforts to organize a polar expedition in search of a northern sea route, and included a letter he asked Frankin to forward to St. Petersburg. Unfortunately, Lomonosov died before the letter could make its way to him.[107]

In 1789 Radishchev published a biography of a friend who died when they were students together at the University of Leipzig.

---

[107]   For Ezra Stiles's letter to Franklin, dated 20 February 1765, and Franklin's 5 July 1765 response to Stiles, agreeing to act as a go-between, see *The Papers of Benjamin Franklin*, vol. 12, *January 1 through December 31, 1765*, ed. Leonard W. Laberee (New Haven: Yale University Press, 1968), 71-77, 194-96. In answering Ezra Stiles's query about any knowledge he had of Lomonosov's attempts to chart a northern sea route, which Stiles had apparently read of in a London newspaper, Benjamin Franklin reported the failure of the first Russian venture, but assured Stiles that "Lomonosow[v] will set the Matter right." Lomonosov died before Franklin composed his letter; indeed, he died only one month before the first of two unsuccessful polar voyages was launched under Admiral Chichagov (the second effort was undertaken in early 1766). The Stiles missive to Lomonosov, which was to be passed on by Franklin—though that was never actually sent—was published in *American Philosophical Society, Franklin Papers*, vol. 44, 19. Henry Leicester's article, "Znakomstvo uchenykh Severnoi Ameriki kolonial'nogo perioda s rabotami M. V. Lomonosova i Peterburgskoi Akademii nauk," *Voprosy istorii estestvoznaniia i tekhniki*, no. 12 (1962): 142-47, exploring the diffusion of "Russian" science in the North American colonies during the eighteenth century, translates from the Latin and reprints the letter from Stiles to Lomonosov. See also Dvoichenko-Markoff[v], "Benjamin Franklin, the American Philosophical Society, and the Russian Academy of Science," 250-51, for more on Stiles, Franklin, and their near association with Lomonosov.

Introducing a new element in Russian memoir writing, this work, *The Life of Fedor Vasil'evich Ushakov* (*Zhitie Fedora Vasil'evicha Ushakova*) assayed the life of an obscure individual.[108] Radishchev offered his schoolmate's life, which he portrayed as one of unblemished valor, as a sublime example of human behavior and potential. Iurii Lotman, whose writings on Radishchev are consistently thought-provoking, emphasized that Radishchev "regarded heroic behavior on the part of the individual as of enormous significance as it provided an educative spectacle for his fellow citizens, for he constantly reiterated that man is an imitative animal."[109]

In successive representations of his life by admiring contemporaries, Lomonosov was held up to his countrymen as a model, albeit one so heroic that it was almost impossibe to emulate him. Radishchev wrote against too easy an adulation of Lomonosov, persuasively stating in the "Discourse" that posterity would be better served by "not trying to ascribe him an honor for what he did not do or for that on which he produced no effect; we

---

[108]   Radishchev, *PSS*, vol. 1, 153-212. The publication of *The Life (zhitie) of Fedor Vasil'evich Ushakov* induced some dismay within the literary establishment. Princess Dashkova, president of the Russian Academy and sister of Radishchev's friend and patron Aleksandr Vorontsov, thought that focusing on the life of an obscure person could be "dangerous in the times in which we lived." Her brother dismissed the book as "merely useless, since the man whom it was about, Ushakov, had never said anything remarkable, and that was the end of it." *The Memoirs of Princess Dashkova*, ed. and trans. Kyril Fitzlyon, intro. Jehanne M. Gheith, afterward. A. Woronzoff-Dashkoff (Durham, NC: Duke University Press, 1995), 236. *Zhitie* was the usual designation given to written saint's lives. It was also, however, appended to many memoirs in the eighteenth century. See Jones, "Biography," 71-79; and also *Svodnyi katalog*, vols. 1-6. Although Radishchev may well have intended the employment of *zhitie* as a statement, as Andrew Kahn argues in "Self and Sensibility," 46, his usage might simply reflect, however, the fact that there was not yet a fixed term for biography.

[109]   Lotman, "Poetics of Everyday Behavior." 248. For an analysis of Radishchev's *Ushakov* that would lend some credence to Dashkova's fears about its troublesome implications, see Lotman's "Otrazhenie etiki i taktiki revoliutsionnoi bor'by v russkoi literature kontsa XVIII veka," in Iu. M. Lotman, *O russkoi literature: stat'i i issledovaniia (1958-1993)* (St. Petersburg, 1997), 223-26.

will not let blind admiration or prejudice lead us into unreasonable praise."[110] But his overall assessment, encapsulated in his query "Is Bacon of Verulam not worthy to be remembered because he could only show how to advance learning?," is hardly contemptuous of Lomonosov. It has, nonetheless, been poorly integrated into a historiography so inimical to divergent viewpoints. During the nineteenth century the concept of scientific genius as superseding method would bring increasing disrepute to Baconian associations,[111] but at the time the "Discourse" was composed, Radishchev's reference could certainly be construed as flattering. Even if Lomonosov did not merit mention alongside the greatest scientific names, he would serve as a fertile exemplar for those among later generations of Russians who might achieve true distinction—there was no need to exaggerate his attainments. This was an opinion that attracted few followers.

It was Murav'ev's insistence throughout his study on the trailblazing character of Lomonosov's scientific efforts, which fit firmly within and greatly added to the growing myth of Lomonosov as the father of Russian science, rather than Radishchev's more circumspect verdict, which proved decisive in the historiography. Murav'ev's standing alone did not marginalize Radishchev's later, more critical views of Lomonosov, for Radishchev was, as we have seen, working singularly in opposition to a developing hagiography; it did, however, assuredly assist in securing the preeminence of Murav'ev's judgments and those of his analogues.

With Mikhail Lomonosov, of concern here is the way his activities or maneuverings, his conflicts as well as his presumed deeds, at the St. Petersburg Academy of Sciences contributed to his evolving identity as a "man of science"[112] during the time of

---

[110]   Radishchev, *Journey*, 236-37.

[111]   Yeo, "Images of Newton," 257-61; Higgitt, *Recreating Newton*, 47-9, 64-7.

[112]   Steven Shapin, "The Image of the Man of Science," in *The Cambridge History of Science, volume 4: Eighteenth-Century Science*, ed. Roy Porter (Cambridge: Cambridge University Press, 2003), 159-83, cautions against the dilemma evident in much of the historiography of science: the need to force unity and consistency onto the lives of the early-modern natural philosopher.

Elizabeth Petrovna.[113] Given his social distance from the court, Lomonosov's association with it was, of course, mediated by patrons. He exploited patronage to formulate a distinctive (given the time and place) identity for himself. In seeking his own glory, Lomonosov contributed to defining the significance of the "new sciences," as seen solely through the status of the Academy of Sciences, to the monarchy. Throughout his career, Lomonosov's engagement with "scientific problems" was accompanied by a persistent concentration on his reputation.

Lomonosov was the proverbial first "native" Russian scientist to be made a full member of the Academy. That fact, however persistent it was as a later Russian-nationalist trope in the historiography, offered to rationalize both Lomonosov's incessant conflicts at the Academy of Sciences and his near absence in "western" accounts of eighteenth-century science, was determinate less of his social status at the time than the fact that the role of a natural philosopher, even more so a chemist, in eighteenth-century Russia possessed markedly low prestige.

As the dimensions of Lomonosov's biography grew steadily more grandiose over subsequent decades, he came to be portrayed not merely as a scientist in the mold of a Newton, Copernicus, Galileo, or Franklin, but as one whose contributions to the cultural life of his own land were as pioneering as his contributions to science.

---

113   Konstantin Pisarenko, *Povsednevnaia zhizn': Russkogo dvora v tsarstvovanie Elizavety Petrovny* (Moscow, 2003), is a lavishly detailed account of mid-eighteenth-century Russian court life, which contains little systematic analysis of how the court itself functioned. O. G. Ageeva's *Imperatorskoi dvor Rossii: 1700-1796* (Moscow, 2008), though covering the Elizabethan court only briefly (see pp. 127-151), supplements Pisarenko's volume with a bureaucratic study of the court. Pisarenko's recent *Elizaveta Petrovna* (Moscow, 2008), though more descriptive than analytic, usefully supplements his above work on court life, and contains much material—if all of it familiar—on Lomonosov (and the Academy of Sciences). Even if stripped of the quasi-Marxist cant that characterized Soviet-era studies of Lomonosov, Pisarenko does not challenge the prevailing image of Lomonosov as a valiant fighter for "Russian science" whose profound hypotheses and discoveries—uniformly ignored outside of Russia—anticipated work unfairly credited to others.

Lomonosov's resistance to the nearly triumphant Newtonian orthodoxy, which was evident throughout his written and public pronouncements, was glossed over, rewritten as somehow Newtonian, or elided completely. The shaping of Lomonosov's identity into that of the representative Russian natural philosopher began, as we have seen, in his own writings, with his contemporary memoirists each contributing particular biographical elements, largely pertaining to his inevitable ascent to the highest planes of learning, to the canonization of the first Russian chemist and physicist.

As is evident from an examination of the earliest biographies, the essence of Lomonosov's science—as opposed to exclusively a reiteration of his saintly qualities—began to be studied as early as in the eighteenth century with Murav'ev's appraisal, but was recast so as not to undermine the prescience and successes expected of a mythic figure. Russian scientists, historians, writers, and literary scholars began to delve ever more deeply into the details of Lomonosov's science as the nineteenth century progressed; this occurred, however, in conjunction with a strengthening of heroic representations of the father of Russian science, not by jettisoning that imagery. This also made any analysis of Lomonosov's actual legacy a difficult proposition. Then again, a disinterested investigation, in view of the iconic status Lomonosov's name had attained, was perhaps impossible.

Chapter 3

LOMONOSOV IN THE AGE OF PUSHKIN

Nineteenth-century textual depictions of Lomonosov cannot be read as other than hagiographic. Portrayals of him as the first Russian scientist, a dauntless investigator of nature's secrets, and as the heroic progenitor of later generations of scientists, are unreservedly reminiscent of the biographies by Staehlin, Verevkin, and Murav'ev, though they are at the same time more layered and present a more complex set of markers than did these predecessors. The years between the late eighteenth-century origins of the Lomonosov mythology and the middle of the nineteenth century (the 1855 Moscow University centennial was an important episode in Lomonosov studies that ushered in new levels of interpretation) are, in terms of both the trajectories of the myth and the cultural contexts that determined its strength, a rather diffuse period. The mythology clearly retained its power over those interested in the place of science in Russian culture, which for nineteenth-century Russia meant nearly the whole of the emergent intelligentsia and professional classes. But a more conscious awareness of its strength, and perhaps constraints, are also evident in literary reactions to it.

Unraveling the images of Lomonosov as the exemplar of a scientist through the early decades of the nineteenth century requires a rather arbitrary selection of what can be seen as particularly significant representations. Eschewed here has been any effort to trace an overt connection between the work of Russian scientists and that of Lomonosov. There have been innumerable labored attempts to establish the outlines of a linear development leading from eighteenth-century scientific developments in Russia, or more

"The Death of Rikhman"

Linocut by N. G. Nagovitsyn, 1958

precisely from Lomonosov's seminal role in laying the foundations for the sciences, to the subsequent substantive progression of many and varied branches of science, including chemistry, physics, geology, metallurgy, geography, and astronomy over the next century.[1] Attempting this task obviously entails a clear acknowledgement of Lomonosov's lasting influence over the research of later scientists. Influence must be in some manner demonstrated, and so the writings of nineteenth-century Russian scientists have been minutely probed for references, however obscure, to their putative forefather.[2] The results have been a series of strained efforts to force a crudely

---

[1]   Attempts to locate an intellectual linkage between Lomonosov and his immediate—as well as rather more distant—successors have a lengthy lineage that antedates Soviet-era historiography. Indeed, it would be no exaggeration to state that nearly every source on Lomonosov that has appeared since the late nineteenth century is distinguished by this methodological approach. Mikhail Sukhomlinov (see introduction) and especially Boris Menshutkin established it as a central tenet in the historiography. The following two institutional and multi-disciplinary surveys of Russian science, both of which quickly established themselves as "definitive," can be held up as the more recent representative studies: N. A. Figurovskii, ed. *Istoriia estestvoznaniia v Rossii*, vol. 1, part 2 (Moscow-Leningrad, 1957); K. V. Ostrovitianov, ed. *Istoriia Akademii nauk SSSR*, vol. 2 (1803-1917) (Moscow-Leningrad, 1964). See also Iu. I. Solov'ev's now standard history of Russian chemistry, *Istoriia khimii v Rossii* (Moscow, 1985), 3-70. Solov'ev is one of the most prolific figures in the Lomonosov industry, and his otherwise fine text not surprisingly contains axioms about Lomonosov similar to those that appear in its predecessors.

[2]   Lomonosov's name abounds in G. S. Vasetskii and S. R. Mikulinskii, eds., *Izbrannye proizvedeniia russkikh estestvoispytatelei pervoi poloviny XIX veka* (Moscow, 1959). When mentioned, however, it is usually in the context of the reissuing of one or another of Lomonosov's already much-cited papers (his electrical researches received the most attention, as they had from earlier scholars), and is usually accompanied by a dearth of fresh analysis. See also Solov'ev and Ushakova, *Otrazhenie estestvennonauchnykh trudov Lomonosova v russkoi literature*, 18-41; Zubov, *Istoriografiia*. For a carefully reasoned break from what was a severely misleading historiography, see Sheptunova's accomplished *Istoriograficheskii analiz rabot po istorii khimii v Rossii*.

teleological and wholly unconvincing model onto Russian science of the times.

Perhaps it is better to speak of the exalted imagery of Lomonosov having inspired subsequent generations, rather than his having had a palpable influence on them.[3] Even so, inspiration is no more easily corroborated than influence. Examining some of the forces that shaped the myth, may, however, provide some answers as to why Lomonosov's biography so permeated Russian cultural discourse on the sciences. Representative texts and other signifiers that operated to sustain and expand Lomonosov's image as a pioneer scientist are readily identified in this period, but it is their implicit and explicit dialogue with one another and with the mythology as they received it that gave his life story continued meaning.

Especially revealing are Alexander Pushkin's responses to Lomonosov. Pushkin's image in Russian culture became all-encompassing by the end of the nineteenth century, and far surpassed in vitality that of Lomonosov. Despite this, the association of Pushkin with Lomonosov provided much strength to the myth of the latter as a scientist. For as resplendent as Pushkin's iconic national status became, and however much he may have eclipsed Lomonosov's stature in literature, he did not and could not detract from specific representations of Lomonosov as a physicist and chemist. Instead Pushkin's reflected splendor only augmented them. The Lomonosov that Pushkin was so motivated by in the 1820s and 1830s was defined both by the heroic eighteenth-century tales and by a handful of fascinating memoirs issued in the first decades of the century. These accounts not only reinforced the pre-existing legends; they also inserted elements that more surely provided for their continued resonance to later generations of Russians.

---

[3]   An argument forwarded but not explored in Alexander Vucinich, *Empire of Knowledge: The Academy of Sciences of the USSR (1917-1970)* (Stanford: Stanford University Press, 1984), 27. See also T. I. Rainov, "Russkoe estestvoznannie vtoroi poloviny XVIII v. i Lomonosov," in *Lomonosov: sbornik statei*, vol. 1, 318-388. Rainov's definition of inspiration is indistinguishable from that of influence.

Vasilii Severgin (1765-1826), a mineralogist, chemist, metallurgist, and educator,[4] was the first Russian scientist to offer a wide-ranging appraisal of Lomonosov as a natural philosopher. Severgin was a leading member of both the Academy of Sciences and the Imperial Russian Academy (an institution devoted to the study of Russian letters and the Russian language). Such outward professional breadth was not unprecedented; indeed, many of the foremost Russian scientists and naturalists of the day, such as Kotel'nikov, Rumovskii, Ozeretskovskii, Protasov, and Lepekhin, were active in both bodies. Of course, this did not so much imply that these were figures of encyclopedic accomplishments across the various arts and sciences, although the fields that they worked in were certainly many and varied, as it underscored the still amorphous boundaries between vocations.

This indeterminacy was a commonplace characterizing both Russian and West European scientific and cultural life from the seventeenth to the early nineteenth centuries, and encouraged the variegated activities of natural philosophers, who adeptly played an array of seemingly disparate roles in society.[5] The continuing absence of a sharp demarcation between, for example, science and literature is profoundly reflected in the types of writings through which Lomonosov's legacy as the father of Russian science was

---

[4]   Sukhomlinov, *Istoriia Rossiiskoi Akademii*, vol. 4, 6-185. A more recent, and comprehensive, study of Severgin is N. N. Ushakova and N. A. Figurovskii, *Vasilii Mikhailovich Severgin, 1765-1826 gg.* (Moscow, 1981). While their work is overall a judicious biography, Ushakova and Figurovskii lose their restraint when faced with interpreting Severgin's views of Lomonosov, which they present as intensely reverent. They see the fact that Severgin was born in the year of Lomonosov's death as "symbolizing" their shared determination to advance the development of "enlightenment, culture, and the sciences in Russia" (ibid., 5). This unity of Lomonosov and Severgin is pursued at intervals throughout their work.

[5]   For autobiographical and biographical efforts to represent, usually in a strikingly unified narrative, the complex lives of early modern natural philosophers, see Shortland and Yeo, *Telling Lives in Science*; Haynes, *From Faust to Strangelove*, 1-65 (her studies of Bacon and Newton yield the most useful material); Shapin, "The Man of Science"; and Söderqvist, *History and Poetics*.

sustained and passed on to future generations during the first decades of the nineteenth century. Severgin was primarily a scientist, but in his approach toward Lomonosov he examined all the facets of his working life. Simply put, he looked at both the chemist-physicist and at the poet. The focus here will be on disentangling the myth of the scientist from the myth of the littérateur, and attempting to reconstruct the image that Severgin presented of Lomonosov as the archetypal Russian natural philosopher.

In November 1805, Vasilii Severgin stood before a distinguished gathering at the Imperial Russian Academy (among those in attendance were Murav'ev and Gavriil Derzhavin)[6] and delivered a lengthy panegyric (*Pokhval'noe slovo*) to the memory of Mikhail Vasil'evich Lomonosov.[7] He began his speech with an elaborately phrased paean to Lomonosov's larger than life qualities:

> The diffusion of a new light into the sciences, the opening up of new paths of growth that are leading towards the heights of perfection, and leading us along those paths on the first difficult journeys: this is the essence of his deeds, and to be endowed with such great abilities is granted to only the rarest of men.[8]

Manifest in his introduction is what would become the message returned to again and again in his oration: Lomonosov's role in blazing new trails in Russian culture, most remarkably in the sciences, an area with limitless potential, was his best gift to his country.

Even conceding the point that flights of rhetoric are requisite in what was after all a commemorative occasion,[9] that Severgin was

---

[6]　Sukhomlinov, *Istoriia Rossiiskoi Akademii*, 161. Derzhavin was the most respected Russian poet of the day.

[7]　V. M. Severgin, *Slovo pokhval'noe Mikhailu Vasil'evichu Lomonosovu* (St. Petersburg, 1805). The printed version, which apparently does not deviate in substance from his speech, is fifty-five pages in length.

[8]　Ibid., 1.

[9]　Why a eulogy was offered to Lomonosov at this particular juncture, forty years after his death, is unclear: the records do not furnish any reasons for it. Alexander Vucinich suggests that Severgin's address represented, for

in thrall of Lomonosov's image cannot be doubted. His panegyric or eulogy,[10] which to present-day readers would better be classified as a biography, captures the core of the mythology forged by his precursors, but at the same time the author adds his own particular luster. Public eulogies to scientists in the eighteenth and early nineteenth centuries served as the primary method by which science was "popularized."[11] It was certainly a principal means for the diffusion of knowledge about Lomonosov in Russia.[12]

---

the speaker and the scientific community, "the triumph of the 'Russian' orientation in the Academy and made it possible to rectify a grave omission of the earlier era: the presentation of some kind of encomium to Lomonosov." Vucinich, *Empire of Knowledge*, 38. The author alludes, of course, to the absence of what was believed to have been a proper eulogy to Lomonosov after his death. His broader claim, that Severgin's speech signified an organized Russian national campaign of some sort, is unsupportable. At times Vucinich adheres too closely to historians such as Sukhomlinov and Pekarskii who, working at a time when the Academy's foreign orientation was a heated issue, unfortunately though perhaps inevitably projected their own preoccupations back onto earlier developments in Russian science.

[10] Panegyrics and eulogies are a complex literary form and should not, in the case of encomiums delivered at various scientific academies in the period under discussion, be viewed "simply as a collection of exemplars of the figure of the ideal natural philosopher, but as an arena in which different explanations of the kind of power at the disposal of the natural philosopher untidily contend. The loosely related form of eulogy, panegyric, and hagiography are all concerned with the chemistry of moral authority." Outram, "The Language of Natural Power," 153.

[11] See Paul, *Science and Immortality* (Paul's work focuses on the Paris Academy of Sciences, where the eulogistic tradition was most developed).

[12] All of the eighteenth-century biographies of Lomonosov could, of course, be just as easily labeled encomiums or panegyrics. Later Russian representations of Leonhard Euler, which were always fulsomely admiring, seem also to have been structured to a degree by the eulogy delivered to his memory at the Academy of Sciences in 1783. Euler's eulogy was given by Nicolas Fuss, the permanent secretary of the Academy, and a former student of his (Fuss married into the Euler family). See Nicolas Fuss, "Eloge de Monsieur Léonard Euler, lu à l'Académie impériale des sciences de S.-Pétersbourg dans son assemblée du 23 octobre 1783 par M. Nicolas Fuss," *Nova Acta Academiae scientiarum imperialis Petropolitanae* 1 (St. Petersburg, 1783): 159-212. A few weeks prior to Fuss's appearance Jacob von Staehlin had delivered a brief speech in honor of the late Euler at an Academy assembly.

Before assessing Lomonosov's activities as a professor at the Academy of Sciences, Severgin offered his listeners a rather stirring, if also by now quite familiar, biographical sketch of his subject up to the point when he returned to St. Petersburg from his sojourn abroad in 1741.[13] Scientific biographies remained dependent on the notion of genius in which a necessarily spectacular level of childhood acuity foreshadowed the prospective scientist's later eminence.[14] Relying to a great degree on the works of Staehlin and Verevkin— as was demonstrated earlier, they can quite easily be interpreted as a single account—Severgin's narration of Lomonosov's early years reiterates what these biographies presented of the young *pomor*'s perspicacious boyhood. Details from Novikov's essay are also present, though these are far fewer than those which are borrowed from Staehlin and Verevkin's tracts.

Lomonosov's struggles in adolescence and young adulthood to overcome myriad social and material obstacles in his quest for knowledge are poignantly and repeatedly highlighted. The young boy's immersion in those famous introductions to the sciences and literature, Magnitskii's *Arithmetic* and Smotritskii's *Grammar*, are given the obligatory prominence. His emergence from the periphery of Russian civilization is, as always, held up with undertones of amazement. The more momentous junctures marking Lomonosov's celebrated ascent, the Slavo-Greco-Latin Academy; Christian Wolff; his initial forays into chemistry, physics, mathematics, metallurgy, and other sciences; encumbrances overcome; and most importantly, his zealous and creative drive to "elevate" himself, were all eloquently presented to the audience.

Severgin utilized both Lomonosov's autobiographical letters to Ivan Shuvalov and the reminiscences of Lomonosov's life in the North that were compiled by the naturalist Nikolai Ozeretskovskii (and published by him in 1805) in creating his work.[15] He referred

---

13   Severgin, *Slovo pokhval'noe Lomonosovu*, 2-14.

14   For an additional comparative eulogy, see Paolo Frisi's *Elogio* (1778) to Newton (reproduced in Hall, *Isaac Newton*, 108-73).

15   These were Gur'ev, "Izvestie o M. V. Lomonosove," and V. Varfolomeev,

on more than one occasion to Lomonosov's collected works, which, given the nature of the evidence he uses, is quite clearly the 1784-87 edition.[16] The heroic imagery implanted by Lomonosov, and elaborately cultivated by his contemporary memoirists, had plainly lost none of its power to impress.

Surveying the remarkable scale of Lomonosov's contributions to Russian learning, he despaired of his ability to adequately convey its magnitude to his listeners, declaring, "how can I begin to enumerate the deeds of this great man!"[17] Lomonosov's works did after all encompass diverse sciences (physical and chemical observations in particular), literature in its many guises, language, Russian history, and so forth. If organizationally Severgin separated Lomonosov's science from his other occupations, he remained insistent throughout in trying to impress upon his listeners what he perceived to have been the encyclopedic features of Lomonosov's labors, for each of them "testify to the advantages that he brought

---

"Zapiska o M. V. Lomonosove," in *Puteshestviia akademika Ivana a,* part 4, 298-302. Ozeretskovskii went to the Kholmogory region in 1788 and gathered these reports from Stepan Kochnev. They deal very briefly with Lomonosov's early years, as well as providing details on the makeup of his family. Ozeretskovskii also published a short poem, "Verses to a Cup" ("Stikhi na Tuiasok"), apparently authored by Lomonosov in 1734 (see ibid., 303; and also Lomonosov, *PSS,* vol. 8, 7, 864-66). If this date is accurate, it is the first known composition of Lomonosov. Ivan Lepekhin, who Ozeretskovskii honored by compiling the above volume, traveled extensively in the areas around Lomonosov's birthplace in 1771-72 (accompanied for a time by Ozeretskovskii, who was then a student), and penned descriptions of the area. For more on Lepekhin and Ozeretskovskii's journeys in the far north of Russia, see T. A. Lukina, "Ekspeditsii akademika Lepekhina v XVIII v.," *Trudy Instituta istorii estestvoznaniia i tekhniki* 41 (1961): 336-45. Ozeretskovskii and Lepekhin, both of whom rose to membership in the Academy of Sciences, assisted in assembling the 1780s edition of Lomonosov's collected works issued by the Academy. Their interest in Lomonosov would seem to have been a longstanding one.

16   Severgin, *Slovo pokhval'noe Lomonosovu,* 4. Severgin alerts the listener/reader to the biography published with Lomonosov's collected works. This was the Verevkin memoir included in the 1784-87 edition of Lomonosov's papers. Much of his correspondence with Shuvalov was also published in this collection.

17   Ibid., 14.

to the country." So how then "is it even possible to pay tribute to his creations, his zealousness, and his talents? It would be done by honoring his sciences, the glory he brought to the fatherland, and the blessings with which his many varied works are adorned."[18]

After reviewing at imposing length Lomonosov's achievements in Russian letters,[19] over the course of which he reprinted sizable excerpts from some of his best-known endeavors in poetry (the area in which Lomonosov made his greatest impact on literature) Severgin moved his focus to the sciences. In making this shift, he begged his listener's continued indulgence: "Although I have already overburdened your attention, esteemed listeners, I have an obligation to show to you the gentleman's other areas of exercises."[20] He strongly insinuated that it was in the sciences that Lomonosov made his most formidable strides forward. The readers did not have to be reminded that Lomonosov was, after all, initially "dispatched to foreign lands for the study of experimental physics and chemistry. And in these sciences he showed himself to be no less useful to the fatherland, he showed himself possessed of no less knowledge, he showed himself in no way less industrious," than he was in his many other activities. Special note was made of the fact that "Lomonosov reformed [in fact built] and enriched the Academy's Laboratory, and did so in accordance with the chemical knowledge of the day, and in the above facilities [he] carried out a great number of chemical and physical experiments."

Due to continuing official neglect, Russian scientists were only fitfully able to engage in laboratory research in their own country after their return from Western Europe, where most Russian chemists, physicists, and other scientists still received their advanced, and in many cases basic, instruction.[21] Study and research

---

[18]   Ibid., 15.

[19]   Ibid., 15-40.

[20]   Ibid., 41.

[21]   Brooks, "Formation of a Community of Chemists in Russia." Brooks's study of the emerging chemistry profession applies with a high degree of equivalence to would-be physicists as well. Severgin himself studied for

abroad (*komandirovki*) remained a fundamental feature on the career paths of most Russian scientists until the early years of the Soviet Union. Severgin's use of Lomonosov's struggles and successes, might offer some succor to those in the still very nascent Russian scientific community interested in expanding opportunities, or simply facilities, in Russia. The employment of mythical forebears generally could be expected to accompany efforts to raise the status of science and of the scientific practitioner.[22]

Appropriating a wonderfully reflective, and self-fashioning, remark that Lomonosov made to Shuvalov,[23] Severgin re-affirmed that for Lomonosov, his exertions in physics and chemistry "served him more as a means for relaxation, than they were ever an arduous form of toil."[24] This was a subtle remonstration of sorts by him to any people present who were under the assumption that Lomonosov was a poet forced by the nature of his position at the Academy of Sciences to engage in scientific work. The reverse was in fact the case. Lomonosov was compelled by the demands of patronage and the need to advance his standing in the cultural hierarchy, where the role of the natural philosopher was as yet weak, to squander his valuable time in composing odes to his masters, both those close at hand and those at the court. By this Severgin does not deny the great benefits that Lomonosov accrued from assiduously cultivating his patrons—he singled out Shuvalov and Mikhail Vorontsov—but it was, perhaps, in the sciences where he truly honored them.[25]

---

some years under Johan Gmelin at Goettingen University. See Ushakova and Figurovskii, *Severgin*, 22-23.

22   Biagioli, *Galilio, Courtier*, 87-88; Cantor, "The Scientist as Hero," 172; Outram, "Scientific Biography and the Case of George Cuvier," 102.

23   *Polnoe sobranie sochinenii Lomonosova*, vol. 1, 1784, 323 (see also Lomonosov, *PSS*, vol. 10, 475). This letter was quoted at length in Chapter 1 (see footnote 28).

24   Severgin, *Slovo pokhval'noe Lomonosovu*, 41.

25   Ibid., 47-55. Unlike Radishchev, that dissenting voice in the early mythology around Lomonosov, Severgin saw, accurately, considerable privileges accruing to Lomonosov from a successful manipulation of patronage.

Looking over Lomonosov's more theoretical physical and chemical researches; Severgin nodded approvingly towards several of those published in the eighteenth century by the Academy's journal, "*Kommentarii*" (*Commentarii*, and later *Novi Commentarii*).[26] These papers clearly demonstrate that "as an experimenter he evinced himself to be of a curious and energetic nature." They were, however, as he stressed, published in Latin, making them, it would seem, less useful as signposts for future, or even contemporary, scholars. Severgin did not deal with the substantive content of Lomonosov's disquisitions; therefore a suspicion that he viewed them as anachronisms is difficult to avoid.

Later in his speech Severgin confirms this suspicion. In a suggestive redirection, he leads the audience's attention beyond those dissertations toward others that he argues are of "greater benefit to our country, those that have been published in the Russian language."[27] By specifying Russian-language works he was limiting the discussion to the less linguistically and scientifically foreboding papers, those that might consequently serve more easily as emulative objects for scientists and would-be scientists. Additionally, though, and more fundamentally, his concern was with writings of decidedly greater "practical" content: precisely those works that offered a certain level of accessibility, along with the promise of potential profit for the country, more than that which was afforded in Lomonosov's corpuscular excursions.

Among the papers, all published or publicly read in Lomonosov's lifetime,[28] recalled by Severgin as particularly useful products of Lomonosov's "fecund scientific mind," are *Oration on the Usefulness of Chemistry*; *Discourse about Air Phenomena, Caused by Electricity*; *Oration on the Origins of Light, Representing a New Theory of*

---

[26]   Ibid., 41-42.

[27]   Ibid., 42.

[28]   On the extensive dissemination of Lomonosov's scientific works in his lifetime, see D. V. Tiulichev, *Knigoizdatel'skaia deiatel'nost' Peterburgskoi Akademii nauk i M. V. Lomonosov* (Leningrad, 1988), 213-76; *Svodnyi katalog*, vol. 2, 162-77.

Colors; A Discourse on the Birth of Metals from the Quaking of the Earth; Discourse on Greater Accuracy of the Sea Route; The Appearance of Venus Before the Sun.[29] Severgin's attenuated explications are repetitive, attempting primarily to impart the speaker's awed reaction to the profundity of Lomonosov's general knowledge and analyses. Each of these papers was either directly considered or at least alluded to by one of the eighteenth-century memoirists. Nevertheless, because they so unequivocally echo tropes in the mythology, some of Severgin's assertions do warrant further perusal.

Since the memory of the death of Georg Richmann while experimenting with a thunder machine was apparently still vivid in the historical memory of those gathered, Lomonosov's work with him on electrical researches was given an admiring testimonial. In this area "is found the absolute reflection of his belief in detailed investigations, a full understanding of the physical and chemical knowledge of the day, and even," in an aside pertaining to the dangers that Richmann, and hence he, had faced, "the fearlessness of his experimentation."[30] Lomonosov the intrepid, curious scientist is underlined here, without, however, any imprudent inferences proposed as to the ultimate significance of his findings.

A presumed association with Benjamin Franklin, much less a positing of Lomonosov's anticipation of his electrical hypotheses, is not present in Severgin's rendering of the issue. Franklin's name was long highly regarded in Russian cultural and scientific circles.[31]

---

[29]   Severgin, *Slovo pokhval'noe Lomonosovu*, 42-44.

[30]   Ibid., 42-43.

[31]   Franklin's reputation assumed a more solely political hue as the nineteenth century progressed, usually at the expense of his scientific life. As we observe in Radishchev's *Journey* (p. 236), however, Franklin's politics were already significant in the first Russian representations of him. By the Soviet period, he was cast as a revolutionary whose social and scientific views were united in a "progressive" deistic worldview. The apotheosis of this approach occurred during the 1956 Franklin Jubilee held at, among other venues, the Academy of Sciences and Moscow University. The idea that Lomonosov was Russia's Franklin—albeit that as a pure scientist he was superior to Franklin, and that it would be better to say that Franklin his country's Lomonosov, was also presented as self-evident. Despite occasionally overstated rhetoric,

Comparisons between him and Lomonosov were freely proffered in
the earlier memoirs (though Murav'ev tenaciously advocated their
equivalence), as was discussed previously, and they would appear
frequently in later evaluations of him as well.[32] In Severgin's address

---

serious care was still accorded to Franklin's science, as for example in Petr
Kapitsa's speech given during the main Franklin ceremony at Moscow
University: "Nauchnaia deiatel'nost' V. Franklina," *Vestnik Akademii nauk
SSSR*, no. 2 (1956): 65-75. Two hundred years of Russian "interest" in
Franklin is dealt with in M. I. Radovskii, *Veniamin Franklin i ego sviazi s
Rossiei* (Moscow-Leningrad, 1958). Unfortunately, Radovskii's inability to
utilize western documentary collections severely restricted the scope of his
work. I. Bernard Cohen treats the fame that Franklin enjoyed in Western
Europe and Russia that arose due to his electrical research in *Benjamin
Franklin's Science* (Cambridge, MA: Harvard University Press, 1990), 112-
14. See also Eufrosina Dvoichenko-Markoff[v], "Benjamin Franklin, the
American Philosophical Society, and the Russian Academy of Science,"
Proceedings of the American Philosophical Society 91, no. 3 (1947): 250-51, for
more on Franklin and his aborted association with Lomonosov. W. Chapin
Huntington's "Michael Lomonosov and Benjamin Franklin: Two Self-Made
Men of the Eighteenth Century," *Russian Review*, vol. 18, no. 4 (October
1959): 294-306, would be of greater interest if the author had actually
attempted a comparison of the two men. Sue Ann Prince's *The Princess & The
Patriot: Ekaterina Dashkova, Benjamin Franklin, and the Age of Enlightenment*
(Philadelphia: American Philosophical Society, 2006), offers little that is new
on either Franklin or his associations with or interest in Russia.

[32]   Until the early years of the nineteenth century, however, when Menshutkin
began to issue the first of his many works on Lomonosov, the associations
between Franklin and Lomonosov were posed in much the same manner
as the eighteenth-century accounts: assumptions of equivalence—though
decidedly not Lomonosov's either theoretical or experimental superiority—
were postulated, but little sustained discussion was submitted to support
such contentions. D. M. Perevoshchikov (a one-time professor of astronomy
at Moscow University, later an academician), was perhaps the most active
scholar in the first half of the nineteenth century in working to underline
the similar research, if not always conclusions, on electricity of Franklin
and Lomonosov. See for example D. M. Perevoshchikov, "Rassmotrenie
Lomonosova razsuzhdeniia: 'o iavleniiakh vozdushnykh, ot eliktricheskoi
sily proizkhodiashchikh'" (a speech given by Perevoschchikov at Moscow
University in 1831 on Lomonosov's best known electrical paper), *Teleskop*,
no. 4 (1831): 491-500; and *Rukovodstvo k opytnoi fizike* (Moscow, 1833),
423-25, 440-41. Zubov diligently unearthed Perevoshchikov's concern with
Lomonosov, and rewarded it lavish attention in his *Istoriografiia*, 409-24.
The delay in popularizing this research seems to be due to the fact that
Perevoshchikov was a scientist who made an occasional reference, albeit

such an overt correlation was averted. In this, and much else, there are here indirect reverberations of Radishchev's "Discourse on Lomonosov."[33] While Severgin was, unlike Radishchev, a scientist, and thus in a better position to assess Lomonosov's work, their evaluations of his originality in areas of natural philosophy parallel each other in questioning certain similar assumptions of the myth. At the same time, both leave considerable room for its further development.

Commenting on Lomonosov's work in optics, which Murav'ev had contrasted favorably with that of Newton, Severgin simply states with reference to Lomonosov's ideas, which he does not furnish exegeses of, that "although they are not in agreement with contemporary notions on the subject, they show his sharp intelligence, and the spirit he brought when striving to conduct investigations."[34] An outright identification with Newton may not have been made, but as a model researcher Lomonosov was, again, still very worthy of being followed. With the exception of Murav'ev's admiring meditation on Lomonosov, distinct juxtapositions of Newton's and Lomonosov's achievements, as opposed to their names,[35] would not become detectable in the literature until Menshutkin at the earliest, writing a century after Severgin's lecture.

For Severgin, Lomonosov's theories did not stand the test of time; yet Severgin does not fail to credit his subject with an amazing array of scientific skills. Lomonosov's study of the earth's geological processes—an area that drew Radishchev's approval—is a case in

---

almost entirely perfunctory, to Lomonosov in his writings. This is not to deny that he had an interest in Lomonosov; he clearly did, but he simply seems to have left little impact on the historiography—or the mythology—until Zubov discovered him.

33   Severgin's knowledge of Radishchev's "Discourse on Lomonosov" might be presumed, but cannot at this time be verified.

34   Severgin, Slovo pokhval'noe Lomonosovu, 43.

35   At least partial confirmation of this can be found in a perusal of Vasetskii and Mikulinskii, Izbrannye proizvedeniia. Analogies made between Lomonosov and Newton in nineteenth-century Russian literature, as opposed to those located in scientific treatises, might yield a different conclusion.

point. While averring that Lomonosov's propositions "are not widely accepted" by contemporary thinkers, he nonetheless insists that they are "respected by the most famous writers working in chemistry, metallurgy, and mineralogy."[36] There is no contradiction in this, for perceptible in Severgin's judgment is the notion that scientific knowledge had presumably progressed since Lomonosov's time, and his ideas, however exceptional for their era, were simply no longer relevant.[37] Such linear thinking has persistently plagued the writing of the history of science, particularly the genre of scientific biography.[38] Its effect on representations of Lomonosov was to keep the substance of his work, at least for a time, at a distant remove from his life.

Advances in navigation, given their centrality to Russian economic and political well-being, not surprisingly invited Severgin's notice. Lomonosov's pains to chart a northern route to the "East" were endorsed, though more for having "proven his knowledge not only in physics, but in mathematics"[39] than for its feasibility. Lomonosov's assumed abilities in physics and mathematics also aided him in "assiduously" conducting observations from St. Petersburg of the passage of Venus before the Sun.[40] The bitter dispute he had with Franz Aepinus over the eclipse, more specifically over who would oversee the observations (Aepinus was technically in charge for a time, which left Lomonosov apoplectic) was, of

---

[36]   Severgin, *Slovo pokhval'noe Lomonosovu*, 43.

[37]   The anonymous author of "O fizicheskikh sochineniiakh Lomonosova," *Atenei*, no. 2 (1829): 110, criticized Severgin's eulogy for overlooking the import of Lomonosov's physical dissertations. Notwithstanding this disavowal, the author, who is widely conjectured to have been Perevoshchikov, based his discussion on the same papers that Severgin mentioned in his speech, and added little to his precursor's conclusions.

[38]   I again refer interested readers to the collection by Shortland and Yeo, *Telling Lives in Science*, where this issue is raised, at least implicitly, in every article.

[39]   Severgin, *Slovo pokhval'noe Lomonosovu*, 43.

[40]   Ibid., 43-44.

course, left undiscussed.[41] Lomonosov's personality often severely undercut his effectiveness, but neither scientific biographies nor mythologies were yet expansive enough to incorporate these less saintly qualities into his written life.[42] Severgin expounded on him in a rousing laudatory assertion that he "was active everywhere,... in everything brought benefits to his country, and ... everywhere was praised for his great worth."[43]

Although "all of the above noted exploits were sufficient to sustain the glory of the great man," Severgin nevertheless went on to cite Lomonosov's interests in metallurgy—this was the area closest to his own work (most of Severgin's writings lay in this field)[44]—to further apotheosize Lomonosov. Scarcely a handbook existed in the Russian language for the different sciences related to mining before Lomonosov, a weakness wholly rectified with his *First Principles of Metallurgy or Mining;* which has proven to be a guide of profound value to Russia.[45] This was not only a work of tremendous practical use, but in its essence it confirmed that the author "was not only the first among Russians, but also an inventive chemist and metallurgist."[46] Given Severgin's own vocation(s), this claim by him carried particular weight. His assessment may have

---

[41]   For documents testifying to Aepinus's and Lomonosov's rancorous arguments over the eclipse, see Pekarskii, *Istoriia Akademii nauk,* vol. 2, 730-34.

[42]   The controversy that ensued after J. B. Biot, in an entry in the *Biographie Universelle* (1822), forcefully put the question of Newton's apparent mental breakdown into the historical discussions (see Hall, *Isaac Newton,* 180-81; Yeo, "Images of Newton," 274-75) does not yet seem to have fully abated. Frank Manuel's *Isaac Newton* is an extended exercise in trying to match a heterogeneous personality to Newton's biography. Whether or not Manuel's product approximated Newton's personality is debatable.

[43]   Severgin, *Slovo pokhval'noe Lomonosovu,* 44.

[44]   Sukhomlinov, *Istoriia Rossiiskoi Akademii,* vol. 4, 339-95 (Sukhomlinov provides a fairly complete index of Severgin's writings). See also Uskhakova and Figurovskii, *Severgin,* 67-110.

[45]   For Severgin's excited reaction to Lomonosov's mining and metallurgical guide, see *Slovo pokhval'noe Lomonosovu,* 44-46.

[46]   Ibid., 46.

been without evaluative meaning, but it nonetheless matched the needs of the genre perfectly.

Undoubtedly the discerning listener came away from the assemblage with the impression that Severgin had consistently emphasized sciences that were, to repeat the point, of immediate economic use to the Russian realm. He did stress those areas, and strongly, but he also was plainly cognizant of the need to pursue theoretical research in Russia, and once again used Lomonosov's image as the personification of this indispensable combination of talents. The speaker closed his examination by accenting an essential aspect of an eighteenth-century scientist's makeup, one already embedded into Lomonosov's biography by Staehlin and his contemporaries, albeit without the scientific "authority" of Severgin:

> There are two characteristics that we see only imperfectly combined in men of science, particularly in the physical sciences in which he was most active. Either one is a good theorist, and a poor practitioner, or one is good practitioner, but a poor theorist. Lomonosov was able to reverse this, for he was very intuitive in his speculations, and with great success he labored with his hands.[47]

To make this supposition, so fundamental to the historiography, unambiguously clear to those notables in attendance at the Russian Academy, for his last illustration of Lomonosov's science Severgin turned to his work on mosaics. Not only did this work reveal his "keen understanding" of abstract concepts in the areas of colors and glass, but it also, in the mosaic art created, it produced a tangible result (Severgin was especially delighted by Lomonosov's *The Battle of Poltava*).[48]

Severgin intimately tied Lomonosov's name to the past diffusion and the future course of the sciences in Russia. Staehlin's

---

[47]   Ibid.

[48]   Ibid., 46-47. Severgin's description of Lomonosov's mosaic endeavors is taken from Staehlin's "Cherty i anekdoty."

famous anecdote concerning Lomonosov's deathbed lament as to what state the sciences in Russia would be reduced to without his guiding hand was submitted to illustrate the argument.[49] Lomonosov's ostensibly "heated championing" of his former student Nikolai Popovskii (who had incurred the wrath of the Holy Synod for his translation of Pope's *Essay on Man*) before his "Maecenas" (Shuvalov) was well-used by Severgin to fortify the image of Lomonosov's determination to advance knowledge of the sciences among the Russians.[50] Imagery left by the eighteen-century mythmakers infuses Severgin's biography. But with his lofty standing in the embryonic scientific community, his greater attention to the breadth of Lomonosov's interests, and his articulated determination to carry on the traditions begun by his "predecessor," Severgin proved himself to be a powerful additional catalyst to the continued relevance of Lomonosov's name in Russian culture.

In an utterance frequently reiterated in the literature, Sukhomlinov maintained that "generations of Russian scientists from Lomonosov to Severgin were linked by the guiding principles of their scientific activities and by the literary inheritance [left by Lomonosov], which flowed out of the living conditions of that time and the historical development of Russian education."[51] This is, again, an argument for influence, which he was determined to detect. Sukhomlinov posited Lomonosov as the fulcrum through which the various strands in early Russian science were joined.

---

[49]   Ibid., 50.

[50]   Fearing the implications of Pope's verses, the Synod forced Popovskii to alter his translation. Despite the Church's determination to fix the text, Pope's allusions to gravitation are still recognizable. On this see Boss, *Newton and Russia*, 224-26; B. E. Raikov, *Ocherki po istorii geliotsentricheskogo mirovozzreniia: iz proshlogo russkogo estestvoznaniia*, 2nd ed. (Moscow-Leningrad, 1947), 287-94; and Klein, *Puti kul'turnogo importa*, 288-90. Whether Lomonosov's attempt to enlist Shuvalov as a defender of Popovskii was as ardent as Severgin states it was depends on how loosely the letter sent by Lomonosov to his patron is interpreted. For the full text of the note, which is briefly excerpted by Severgin (*Slovo pokhval'noe Lomonosovu*, 51), please see Biliarskii, *Materialy dlia biografii Lomonosova*, 215-216.

[51]   Sukhomlinov, *Istoriia Rossiiskoi Akademii*, vol. 4, 2.

The mere existence of Severgin's account would seem to highlight perfectly this influence, except that his speech more readily lends itself to a quite different reading. His remarks are filled with an image of Lomonosov, lengthily though vaguely drawn, as the originator of science in Russia, but he does not outline direct bonds between his work and that of later generations of scientists: a crucial distinction. Concomitantly it also demonstrates convincingly the attempts by figures such as Severgin to define more sharply and hence promote the status of science in Russia at the beginning of the nineteenth century.[52]

The writings of Lomonosov's contemporaries and immediate "successors" are the more visible elements in the imagery that emerged in the decades following his death, but they constitute the myth that Pushkin began to acknowledge in the 1820s only in part. Less prone to description is the emotive sway that Lomonosov's name exerted over various figures of that time. Wonderfully evocative is the notion that the mythology encircling Lomonosov's memory was such that his native region, centered near Kholmogory and Arkhangel'sk, had become an object of pilgrimage by various supplicants.[53] That a series of journeys, commencing with Murav'ev's,

---

[52]   Severgin's address came at a time when a "love for science" corresponded to a "love for Russia." Sukhomlinov, cited in Vucinich, *Empire of Knowledge*, 38. Sukhomlinov's point can be accepted without acceding to his broader assertions of an incipient Russian nationalism at the Academy of Sciences. From his study of the eulogies delivered at the Paris Academy in the eighteenth century, C. B. Paul held that what they accomplished more than anything was to "describe the rise of modern science, omit details that would unduly debase the protagonists in the public esteem, and dwell on the new breed of natural philosophers on whose character and investi-gations presumably hang the fate of humanity." Paul, *Science and Immortality*, 109.

[53]   This intriguing idea first came to my attention through Irina Reyfman's study of the origins and evolution of eighteenth-century Russian literary reputations. Reyfman, *Trediakovsky*, 96. She traces its source to the uniqueness of Lomonosov's own biography, for "In myths the benefactor of humankind was supposed to be of obscure origin, and Lomonosov's low social station, once a negative trait, became a sign and precondition of his extraordinary destiny" (ibid., 95). The scientist in eighteenth-century Russia was even less honored than was the humble littérateur, thus the social distinctions among

was undertaken is definite. These pilgrimages seem to have terminated with Pavel Svin'in's in the 1820s. After this time, records of journeys to Lomonosov's former home that resemble spiritual pilgrimages are difficult to find.[54] Leaving aside this fascinating, if less tangible, element of the Lomonosov mythology, each of these travelers (among them Murav'ev, Lepekhin, Chelishchev, and Ozeretskovskii) left behind a report of their visit. Svin'in's journey, and the story he wrote of it, would eventually have a sure impact on Lomonosov's historiography.

In addition to being an editor, writer, artist, and occasional diplomat, Pavel Svin'in (1787-1839) was an avid collector of antiquities from the Russian past.[55] His role in the literature on Lomonosov encompasses both the form of ethereal journey and a more precise form of historical investigation. Svin'in was born in Arkhangel'sk, and his attention to Lomonosov appears to have been of long standing. As early as 1812, while posted on a consular mission to the United States, he presented a short piece

---

natural philosophers was far less severe. This was reflected in a double-edged manner in the early biographies of Lomonosov. The birth in such an obscure locale of this revered figure was hailed as noteworthy, and without obvious negative connotations, yet the very meanness of his background would paradoxically seem to stymie the ascent to respectability of the sciences he was cast as embodying.

[54]   By the 1820s the sentimental travelers, whose journeys were motivated by personal curiosity and structured by their "emotional responses to what they saw," had long since displaced the religious pilgrim as a literary form in Russian letters. See Schönle, *Authenticity and Fiction*, 3. Svin'in's excursion, undertaken to venerate a secular deity, but a deity nonetheless, might be viewed as an anachronism. With its partial mixing of elements, some decidedly religious or mythological and others more prosaic and decidedly secular, it perhaps signified the end of the spiritual journey to Lomonosov's native area.

[55]   There are no detailed works on Svin'in; the most comprehensive is the now-dated V. V. Danilov, "Dedushka russkikh istoricheskikh zhurnalov ('Otechestvennye zapiski' P. P. Svin'ina)," *Istoricheskii vestnik: istoriko-literaturnyi zhurnal* 141 (1915): 109-29. Marina Swoboda and William Benton Whisenhunt, *A Russian Paints America: The Travels of Pavel P. Svin'in, 1811-1813* (Montreal: McGill-Queen's University Press, 2008), includes a brief biography: pp. 24-33.

on Lomonosov, composed by Nikolai Karamzin for his *Pantheon of Russian Authors*, to the American Philosophical Society.[56] The entry on Lomonosov that was translated into English by Svin'in dealt, as the essay's title indicates, with its subject's literary gifts. Karamzin admired Lomonosov's poetry, but it was as an ideal that he might best be remembered: "He put down his name in the book of immortality-there, where glowe [sic] the names of Pindars, Horaces, Rousseaux [sic].... If the Genius & talents of the mind have a right to the gratitude of the people—Russia owes a monument to Lomonossoff [sic]."[57]

Svin'in's testimonial to Lomonosov was his journey to Arkhangel'sk in 1828, which he followed with an article, "Descendants and Contemporaries of Lomonosov" ("Potomki i sovremenniki Lomonosova", 1834),[58] which outlined, in broad strokes, his trek. Admiring exhortations continuously punctuate his narrative. "Without the exploits of Lomonosov, it would scarcely be possible for our own scientists to write in our own language." Indeed, he continues, it would be unlikely that the Russian Academy of Sciences would have any Russian scientists.[59] As always, Lomonosov's Russian language compositions are especially cherished.

What distinguishes Svin'in from the earlier memoirists is not his representation of Lomonosov, but rather that it seems he went to Arkangel'sk for the purpose of meeting with Lomonosov's niece, the quite aged Matrena Evseevna Lopatkina, in order to acquire from

---

[56]   Karamzin's *Panteon rossiiskikh avtorov* (1802) was rendered by Svin'in as *the pantheon of the Russian poets*. Svin'in's gift to the Americans was first published in Eufrosina Dvoichenko-Markov, "The American Philosophical Society and Early Russian-American Relations," *Proceedings of the American Philosophical Society* 94, no. 6 (1950): 595-96. In his translation, Svin'in was faithful to the Russian original, for an example of which see N. M. Karamzin, *Izbrannye sochineniia*, vol. 2 (Moscow-Leningrad, 1964), 168-69.

[57]   Dvoichenko-Markov, "The American Philosophical Society," 596.

[58]   P. P. Svin'in, "Potomki i sovremenniki Lomonosova," *Biblioteka dlia chteniia* 2 (1834): 213-20.

[59]   Ibid., 218.

her a trove of Lomonosov's scientific, literary, and administrative papers:

> My pleasant journey to the homeland of Lomonosov was concluded when I acquired, for a reasonable donation, a bundle of manuscripts, which were mostly written in his [Lomonosov's] own hand, and in the Russian, French, German and Latin languages. Examining these valuable documents served me for the entire journey as an inexhaustible source of surprise and curiosity and it convinced me of the truth of the idea that the kernel of genius consists of diverse pursuits, and this truth has been demonstrated to me by Lomonosov, Newton, Leibnitz, and Walter Scott.[60] And here, in this small notebook, one can see samples of the many activities performed by the great man, from complex speculations in mathematics and natural philosophy, to the writing of sonorous verses; from discourses on mining to projects in other fields, and here there are also official papers.[61]

This passage proved to be frustratingly opaque to later scholars, for Svin'in offered no further explication of what these documents might be.[62]

His essay seems to be a sequel to at least one earlier article, "News about the Newly Discovered Lomonosov Manuscripts" ("Izvestie o vnov' otkrytykh rukopisiakh Lomonosova," 1827).[63]

---

[60]   The Newton association is anticipated, the Leibniz one less so; as for Walter Scott, his historical novels were immensely popular in Russia in the first decades of the nineteenth century. As noted in Damiano Rebekkini, "Russkie istoricheskie romany 30-x godov XIX veka," *Novoe literaturnoe obozrenie*, no. 34 (1998): 418, at least twenty-five distinct translations were made of his novels in the 1820s alone. An historical novelist himself, and one who modeled himself on Scott, Svin'in would greatly value a comparison between Lomonosov and Scott.

[61]   Svin'in, "Potomki Lomonosova," 229-30.

[62]   An exhaustive dissection of the origins of the so-called *Svin'in Collection* (*Svin'inskii sbornik*), which Svin'in sold to the Imperial Russian Academy in 1836 (this institution was absorbed by the Academy of Sciences in 1841), is found in Kuliabko and Beshenkovskii, *Sud'ba biblioteki i arkhiva Lomonosova*, 13-14, 74-89.

[63]   Svin'in published this in the journal he edited. See *Otechestvennye zapiski* 31,

Here he alleges to have come into possession of 500 pages of Lomonosov's scientific papers, mainly in Russian and Latin, some fifty treatises of which he named. The majority of the writings fall under the rubrics of chemistry and physics, including many that overlapped with what Lomonosov termed "physical chemistry."

Svin'in distanced himself from evaluating the papers, except for those close to one of his own areas of expertise, which apparently was mining; as for the remainder, he would leave them for others to evaluate.[64] Nevertheless, he believed that they had great value:

> It is possible that many of the sciences about which, as seen here, Lomonosov wrote have since his time progressed greatly, but let the enlightened world know how this learned man embraced these sciences, which he with both ease and clarity explained in Russian, even to the most difficult speculations; how he entered into battle with the great Newton and Euler; how many of his ideas, which were considered strange in his day, are now in accord with the systems of present-day thinkers.[65]

A passing familiarity with the contents of Lomonosov's writings is conceivably demonstrated by the reference to Newton and Euler, though which disagreements he had with their theories is not explained. With Newton the differences were many—Murav'ev remarked on their differing hypotheses in optics—but with Euler the discord was purely personal. The notion that Lomonosov's writings were viewed as "strange in his day" testifies to what became an enduring theme in the literature, that his work was little understood by his contemporaries. Beyond this, it is impossible to know how closely Svin'in examined the papers.

The provenance and makeup of the 1827 and 1834 manuscripts have been approached very belatedly and hesitantly. They did not

no. 89 (September 1827): 489-94.

[64]   At the end of his 1834 article (Svin'in, "Potomki Lomonosova," 220), he indicated, accurately, that he had in 1828 published two "excerpts" from Lomonosov's mining papers in the *Mining Journal* (*Gornyi zhurnal*).

[65]   Svin'in, "Izvestie o vnov' otkrytykh rukopisiakh Lomonosova," 490.

become the objects of detailed study until Menshutkin so fruitfully exploited them several decades after their hazy procurement by Svin'in. It may be that in fact the *Svin'in Collection* is comprised of a single series of documents. Questions asked of Svin'in as to the veracity of his 1827 article might have led him to issue "Descendants and Contemporaries of Lomonosov" in response. An inveterate traveler, Svin'in published a copious amount of notes describing the places he had been to.[66] The truthfulness of his accounts was the target of some disbelief by his fellow writers and editors.[67] Much of their criticism likely emerged from the literary polemics of the day, for Svin'in was loosely associated with such journalists as Fadei Bulgarin and Nikolai Grech, who were viewed by many literary figures less slavish before the government as being hopelessly reactionary. Some of the mockery, though, may be due to the fact that

---

[66]   See his illustrated recounting of his travels in the northern United States, primarily the Philadelphia area, see Svin'in, *Opyt zhivopisnago puteshestviia po Severnoi Amerike* (St. Petersburg, 1815).

[67]   Suspicion also fell on the historical sources unearthed by Svin'in and employed by him in various studies. The roots of these doubts are difficult to trace; they seem to have been clearly, though also quite nebulously, "in the air" at the time. An early and widely circulated attack on Svin'in's credulity was A. E. Izmailov's fable "The Liar" (*Lgun*, 1824). See *Poliarnaia zvezda, izdannaia A. Bestuzhevym i K. Ryleevym* (Moscow-Leningrad, 1960), 387-90, 960. Pushkin, in addition to greeting Izamailov's tale with approbation (see A. S. Pushkin, *Polnoe sobranie sochinenii*, vol. 14 [Moscow-Leningrad, 1941], 61-62), in 1830 contributed a humorous short piece alluding to Svin'in's tendencies toward presumably wild exaggeration: "Pavlusha [Svin'in] declared that in the home of his parents were a cook-apprentice astronomer, a post-boy historian, and a poultryman who also composed verses better than Lomonosov." "The Little Liar" ("Malen'kii lzhets"), in Pushkin, *PSS*, vol. 11 (Moscow-Leningrad, 1949), 101. See also Danilov, "Dedushka russkikh zhurnalov," 116-18. Pushkin's reproaches towards Svin'in, however, lacked venom, and they remained in sporadic contact, mainly over historical matters, into the 1830s. Svin'in's skills in mixing history and fiction are made vivid by Richard Wortman in his analysis of Svin'in's published portrayal of Nicholas I's coronation (1826). See Wortman, *Scenarios of Power*, vol. 1, 282-95. If done correctly, and Wortman believes that Svin'in was an effective stylist, then accounts of court ceremonies may have proven to be useful politically to hyper-centralized regimes such as Russia's.

Svin'in was indeed prone to embroidering his stories. We have no evidence to support contemporary suspicions of his Arkhangel'sk expedition, but later scholars were less than comfortable with Svin'in's account of his acquisition of the writings.[68] Whatever the case, Svin'in's role as a pilgrim to Lomonosov's birthplace and the discoverer of many of his "forgotten" papers became inscribed in the mythology.

For those unacquainted with Russia, it is well-nigh impossible to convey the deep import attached to Alexander Pushkin's name in his country's life.[69] Appellations such as poet, dramatist, prose writer, and historian—and he excelled in each of these roles—are

---

[68] "It is unfortunate," wrote P. Perevlesskii, that Svin'in "did not take care to explain what was contained in these papers and where they disappeared to." P. Perevlesskii, *Sobranie sochinenii izvestneishikh russkikh pisatelei*, no. 1 (Moscow, 1846), CXXXI. Pekarskii seemed disinclined to believe that Svin'in had actually gone to Kurostrov, one of the places he claimed to have visited on his pilgrimage to the North, for he found his descriptions of the surrounding area thin. See Pekarskii, *Istoriia Akademii nauk*, vol. 2, 277. Perkarskii relies on Svin'in as a source, though he is hesitant about accepting his essays at face value, for he "loved to embellish and exaggerate in his stories" (ibid., 881).

[69] Thoughtful approaches to the origins and evolution of the "cult" of Puskhin in Russian culture include: Paul Debreczeny, *Social Functions of Literature: Alexander Pushkin and Russian Culture* (Stanford: Stanford University Press, 1997); V. M. Esipov, *Pushkin v zerkale mifov* (Moscow, 2006); Boris Gasparov, Robert P. Hughes, and Irina Paperno, eds., *Cultural Mythologies of Russian Modernism: From the Golden Age to the Silver Age* (especially part II, "Pushkin as an Institution") (Berkeley: University of California Press, 1992), 183-250; Marcus C. Levitt, *Russian Literary Politics and the Pushkin Celebration of 1880* (Ithaca, NY: Cornell University Press, 1989); Stephanie Sandler, *Commemorating Pushkin: Russia's Myth of a National Poet* (Stanford: Stanford University Press, 2004); and Victor Terras, "Some Observations on Pushkin's Image in Russian Literature," *Russian Literature* 14 (1983): 296-316. Perhaps the best introduction to the meaning of Pushkin in Russian culture, however, is Abram Tertz's (Andrei Sinyavsky) *Strolls with Puskhin*, trans. Catherine Theimer Nepomnyashchy and Slava I. Yastremski (New Haven: Yale University Press, 1993). Tertz/Sinyavsky's occasionally insolent, and often maddeningly elliptical, attempt to discover Puskhin proved to be very controversial when it was finally released in the Soviet Union in 1989 (it was first published in the West in 1975, soon after the author's emigration). Puskhin's image remains imposing.

unrevealing, and altogether fail to depict his sacral status. Suffice it to say that debates, mainly but not exclusively literary debates, over Pushkin have shaped discussions over the state of Russian culture since his mythologically appropriate death following a duel in 1837. Lomonosov, as the creator of "modern" Russian letters, was a subject of considerable fascination to Puskhin. Subsequently, innumerable scholars have investigated Pushkin's views on Lomonosov's literary legacy.[70] His many evaluations of Lomonosov, which are found throughout his writings, left a lasting impression on both popular and scholarly images of Lomonosov.

Pushkin's commentaries principally concern the poetic and linguistic traditions Lomonosov left behind, but some of his observations have been quite productively used by those interested in Lomonosov's science to keep his memory both alive and relevant. The emphasis is usually on a few statements by Pushkin—mythmaking is commonly averse to nuance—that soon became clichés in the historiography. It is not for the intuitiveness of his perceptions that his judgements have proven so influential, however eloquently they may be posed, but rather it is Pushkin's own mythical image that made his utterances about Lomonosov so much a part of the literature. Indeed, a Russian audience can easily recall many of his remarks on Lomonosov. Pushkin's *Journey from Moscow to Petersburg* (*Puteshestvie iz Moskvy v Petesrburg*, which he composed in the mid-1830s but which was not published until his later editors gave it a title and printed it), is of particular consequence for the succinct appraisal of Lomonosov that it carried. An analysis of selected earlier assessments by Pushkin will allow a fuller illustration of his body of work on Lomonosov.

It was less his brief references to any of the sciences that brought scholars to Pushkin's work on Lomonosov[71] than his

---

[70] The bibliography on Pushkin's consideration of Lomonosov's literary heritage is dauntingly large, and does not concern us here. That said, I found the following to be of general benefit to my thinking on the topic: Reyfman, *Trediakovsky*; Iu. V. Stennik, *Pushkin i russkaia literatura XVIII veka* (St. Petersburg, 1995).

[71] A partial exception is M. P. Alekseev's "Pushkin i nauka ego vremeni," in

articulate configuration of him as a cultural hero that proved so rich
a resource. "'We have criticism, but no literature.' Where did you
get such a notion? It is criticism that we lack. Hence the reputation
of Lomonosov," wrote Pushkin to Aleksandr Bestuzhev in 1825.
It was certainly not for the quality of his writings that Pushkin
remembered him: "I respect him as a great man, but certainly not
a great poet."[72] And it was never Lomonosov's creative output, be
it in poetry, drama, language studies, history, or the sciences, that
ever seemed to exercise Pushkin's passion. Lomonosov's being the
pioneer scientist and poet among the Russians of his time was itself
what made Lomonosov a worthy figure. When judging the actual
products of Lomonosov's imagination, however, a degree of care
was required.

Reacting to Lemonté's short French review of Russian
literature (1825) gave Pushkin the opportunity to more fully assess
Lomonosov's legacy. Devoted to literary issues, his critique crossed
restrictive disciplinary boundaries that were formed later. He
cautions against supposing that each of Lomonsov's callings could
be valued equally:

> Combining uncommon will power with uncommon strength
> of understanding, Lomonosov embraced all the branches
> of knowledge. A hunger for knowledge predominated
> among the many passions that charged his spirit. Historian,
> rhetorician, mechanic, chemist, mineralogist, artist and
> poet, he was experienced and penetrating in all of these....
> He is the first to delve deeply into our country's history,

M. P. Alekseev, *Pushkin: sravnitel'no–istoricheskie issledovaniia*, eds. G. V. Stepanov and V. N. Baskakov (Leningrad, 1984), 22-173. Alekseev's article
deals mainly with poetic expressions of natural philosophy and Pushkin's
responses to them, and only to a degree with Lomonosov; nonetheless,
it is instructive as a reminder that poetry was a crucial medium for the
popular dissemination of science in Russia. Less engaging is Iu. I. Solov'ev,
"M. V. Lomonosov v otsenke A. S. Pushkina," *Voprosy istorii estestvoznaniia
i tekhniki*, no 4 (1983): 65-69, which, notwithstanding Solov'ev's mastery of
the literature, fails to contextualize Pushkin's interest in Lomonosov.

[72] Pushkin, *PSS*, vol. 13 (Moscow-Leningrad, 1937), 177-78. His letter was in
answer to Bestuzhev's "Vzgliad na russkuiu slovesnost' v techenie 1824
i nachale 1825 godov" (1825).

and to establish the rules of its official language, he gave us laws and forms of classical eloquence, and together with the unfortunate Rikhman [sic] he anticipated Franklin's discoveries, established a factory, constructs the machines himself, graces art with his mosaic creations.[73]

The death of Richmann, the similarities to Franklin's hypotheses, the work in mosaic arts, the building of his glass factory, and, always most importantly, the incessant quest for learning—in short, many of the by-then quintessential components of the Lomonosov myth, are revisited by Pushkin. As the earlier memoirists who had made these elements so fundamental to Lomonosov's biography were increasingly forgotten, Pushkin's voice became more and more the referent.

Pushkin concentrated mainly on Lomonosov's poetic gifts, and felt it incumbent on him to remind his readers that:

Poetry is, for the few born to be poets, the single passion that embraces and engulfs all of the attention and all the exertions, and all the impressions of their lives: but when we begin to investigate the life of Lomonosov, we find, that the exact sciences were always the main and favorite of his occupations, and versification though often an amusement, was more often a necessary exercise.[74]

Here he paraphrases one of Lomonosov's best-known letters to Shuvalov, which was, as we have seen, also employed by Severgin. But unlike the myriad others in the literature who cite this letter rather narrowly, either to outline Lomonosov's many-sided genius or to emphasize that he was a scientist forced by his patrons to compose verse, Puskhin clearly appears more interested in writing against the myth of Lomonosov in Russian letters.[75] For to Pushkin

---

[73]   Pushkin, "O predislovii g-na Lemonte k perevodu basen I. A. Krylova," *Moskovskii telegraf,* part 5, no. 17 (1825): 42. I have relied, with some modifications, on Tatiana Wolff's translation of the text (see her *Pushkin on Literature* [Evanston, IL: Northwestern University Press, 1998], 122-23).

[74]   Pushkin, "O predislovii Lemonte," 43; Wolff, *Pushkin,* 123.

[75]   Irina Reyfman reads Pushkin's overall article somewhat differently and considers that he was, at least on linguistic questions, still in Lomonosov's

that necessary "passion" was singularly lacking in Lomonosov's poetry. Pushkin's assertions have moreover been used to bolster that persistent notion, which is arguably true, that Lomonosov was first and foremost a natural philosopher.

Quoting Sumarokov's famous verse on Lomonosov: "He is the Malherbe of our Lands, he is like Pindar,"[76] Pushkin announces what he deems "Lomonosov's true achievement as a poet." Lomonosov would serve well as a symbol for his country's literature, as did Malherbe and Pindar for theirs, but there was no need to bestow false praise on the work itself, for "it is strange to complain that fashionable people do not read Lomonosov, and to expect that a man who died seventy years ago, should still be the public's favorite. As if it is the great Lomonosov's fame stands in need of the trivial honors bestowed on a stylish writer." This last remark spoke not only to Lomonosov's own writings, but also to the role and status of the writer in Russian society, an issue that, given Pushkin's own uncertain standing,[77] profoundly concerned him. Beyond its meaning for Russian literature, the works were no longer "fashionable"—would it also not be absurd, Pushkin reasoned, to expect scientists to still take seriously the products of

---

thrall (*Trediakovky*, 201-02). Iurii Stennik in part evades the question by successive references to Pushkin's otherwise deep respect for Lomonosov (Stennik, *Pushkin i russkaia literatura*, 129-30).

[76]   Pushkin, "O predislovii Lemonte," 43; Wolff, *Pushkin*, 123.

[77]   "Independence and dignity" were critical to Pushkin's self-representation. Humiliatingly dependent on Nicholas I as his personal censor, or more precisely on Nicholas's police chief, A. Kh. Benckendorff, they were not goals he was confident of attaining. Additionally, the development of a wider reading public was altering, and reducing, the power of patronage in Russian literary life during the 1820s and 1830s, which seemed, paradoxically, to broaden the possibilities for Russian writers generally, and to narrow his own. Whether the "gentleman amateur" could mold himself into a "professional man of letters" was, for Pushkin, never resolved. Instead, following his early death, he was transformed into the national bard, blessedly free from such mundane calculations as the marketplace. For an inquiry into Pushkin and patronage, see William Mills Todd III, *Fiction and Society in the Age of Pushkin: Ideology, Institutions, and Narrative* (Cambridge, MA: Harvard University Press, 1986), 51-55, 106-09.

Lomonosov's researches? They too were conducted and written in the "distant past," and serve better as fine historical specimens than working prototypes.

Pushkin gave these ideas a more extensive airing in the entry he drafted on Lomonosov for his unfinished *Journey from Moscow to Petersburg* (1834-35). Through his *Journey from Moscow*, which reads as a polemic with Radishchev's *Journey from Petersburg to Moscow*, Pushkin sought to reintroduce and reinterpret Radishchev to the Russian reading public.[78] Due to censorship restrictions, Radishchev's name and his most renowned work had been removed from public purview since the early 1790s. By criticizing Radishchev's blunt attacks on Russian social conditions, perhaps Pushkin could surreptitiously temper the ban on Radishchev. Despite changes that Pushkin made to the manuscript, his work would not be published until after his death (the revised piece was issued in the 1838-41 edition of Pushkin's collected works).[79] His

---

[78]   Pushkin was as instrumental in fashioning Radishchev's image as he was Lomonosov's. The notion, for example, that Radishchev, acutely despondent at the rebuke he received after proposing a series of legal reforms at the beginning of Alexander I's reign, committed suicide in response, or rather in protest, an act that he had long seen as somehow inevitable for himself, was "popularized" by Pushkin. Radishchev did take his own life, but little can be presumed of his state of mind at that time. See Pushkin's essay "Aleksandr Radishchev" (1836), in Pushkin, *PSS*, vol. 12 (Moscow-Leningrad, 1949), 30-40, 351-56. "Aleksandr Radishchev" was, however, unable to make it past the censorship until 1857. Pushkin was engaged with Radishchev's story most of his adult life, and wrote of him in many disparate pieces. A. G. Tartakovskii's "A. S. Pushkin i A. N. Radishchev: zametki istochnikoveda," *Otechestvennaia istoriia*, nos. 1-2 (January-February 1999): 64-90; (March-April 1999): 142-70, explores Pushkin's study of Radishchev, with an emphasis on his "discovery" of a *Journey from Petersburg to Moscow*. An interesting sidebar is that Pavel Svin'in had, in the course of collecting Russian antiquities, acquired the diary of Catherine the Great's secretary, A. V. Khrapovitskii. He dutifully recorded Catherine's apoplectic reactions to Radishchev's book. Pushkin's concerns with Khrapovitskii's journal, along with his successful negotiations with Svin'in over consulting it, are outlined in ibid., 142-45.

[79]   Copies of both the "original" manuscript and the changes made to it by the author are found in Pushkin, *PSS*, vol. 11, 223-267, 455-94, 562-63 (for Pushkin's assessment of Lomonosov, see especially pp. 230, 248-55, 464).

segment on Lomonosov suffered only minor stylistic emendations during the editorial processes.

As Pushkin's *Journey* has been accepted as disputing Radishchev's *Journey from Petersburg to Moscow*, so has his "Lomonosov" been perceived as a rebuttal of Radishchev's "Discourse on Lomonosov."[80] This is a somewhat blinkered understanding, developing no doubt out of the prevailing supposition that Radishchev's evaluation of Lomonosov was extremely negative. As was suggested earlier, however, his "Discourse on Lomonosov" is better viewed as attempting to delimit the myth of Lomonosov, not as attempting to denigrate it. Any such effort could be interpreted, and indeed has been interpreted, as an attack on Lomonosov.

Therefore, Pushkin's often critical remarks on Radishchev's assessment of Lomonosov, which focus more on the heavy-handedness of his approach than on the substance of what he wrote, might then rather easily be extrapolated into a thesis resting on Pushkin's rejection of Radishchev. Forswearing such an exegesis of Pushkin's *Journey*, Svetlana Evdokimova, in a recent examination of Pushkin's historical conceptions, argues that he "does not outrightly deny the facts presented by Radishchev but reemploys them," or rather, "he tries to correct the astigmatism of Radishchev's historical vision."[81] Doing thus, Pushkin more visibly upheld the myth of

---

[80]   For perhaps the inaugural example, see Pavel Radishchev's defense of his father's "Discourse on Lomonosov" in P. A. Radishchev, "A. N. Radishchev," 432. In a section entitled "On the Abuse of Lomonosov," young Radishchev pointed out that his father's criticisms of Lomonosov were directed first at his tendency to "flatter unworthy idols," and second at the "monotony of his verses." Other than these minor admonitions, Pavel Radishchev emphasized that Pushkin had sharply misunderstood Radishchev's estimation of Lomonosov, which he insisted was overwhelmingly approving.

[81]   Svetlana Evdokimova, *Pushkin's Historical Imagination* (New Haven: Yale University Press, 1999), 89. Evdokimova correctly emphasizes that "Pushkin's *Journey* emerges as a metepoetical work in which the author experiments with consciously constructing a different narrative around the same set of events. And, of course, Pushkin is aware that both narratives— Radishchev's and his own—are fictionalized and subjective" (ibid., 89-90). Evdokimova only peripherally discusses the "Discourse on Lomonosov" in her disquisition (see pp. 93-94), though she applies similar arguments

Lomonosov than Radishchev did, while at the same time broadly adhering to Radishchev's reworking of it.

Pushkin begins a rather enigmatic section by observing that "At the end of his book Radishchev included a passage on Lomonosov," and though he "carefully concealed his intention under cover of respect," he in fact "intended to damage the inviolate fame of Russia's Pindar."[82] Especially "seditious" to Pushkin was the idea that Lomonosov, though himself approaching the pantheon of the great, the "temple of glory" in Radishchev's words, could not himself enter it. But then, Radishchev did query whether "Bacon of Verulam [is] not worthy to be remembered because he could only show how to advance learning?"[83] Further elucidating the Baconian reference, Pushkin complains:

> Radishchev says that in no branch of knowledge did Lomonosov make any new discoveries, and in the same breath compares him to—Lord Bacon! Such were the curious ideas held in the eighteenth century on the greatest thinker of recent times, a man who effected a tremendous revolution in the sciences, putting them on the road which they still follow today.[84]

But as for hailing Lomonosov the "Russian Bacon…. What is the point of such a sobriquet? Lomonosov is the Russian Lomonosov— that serves him well enough." At the time of Pushkin's deliberations, the image of Francis Bacon had been eclipsed by scientific figures

to it. For further research into Pushkin's historical interests, see Andrew Wachtel, *An Obsession with History: Russian Writers Confront the Past* (Stanford: Stanford University Press, 1994), 66-87. Wachtel employs the notion—with an obviously strong suggestion of Bakhtin's formulations—of an "intergeneric dialogue" between texts to elucidate Pushkin's approach.

82   Pushkin, *PSS*, vol. 11, 248. For this segment I have relied on Tatiana Wolff's translation. Please see Wolff, *Pushkin*, 345-46. Wolff translated only part of Pushkin's "Lomonosov." When I employ her rendering of it, it will be noted.

83   Ibid., 346; Pushkin, *PSS*, vol. 11, 248. Here Pushkin is quoting from Radishchev's "Discourse on Lomonosov." (For the full passage, please see Radishchev, *Journey*, 237.)

84   Wolff, *Pushkin*, 349; Pushkin, *PSS*, vol. 11, 230, 464.

that better fit the romantic conception of the hero, with Newton obviously standing preeminently above all others.

The earlier positive associations that accrued to representations of Bacon as the promulgator of the rigorous application of technique, lacking any identifiable discoveries or shattering hypotheses, no longer fit into an era that viewed "scientific genius" as "transcending any simple rules and methods to grasp new laws of nature."[85] A clear assumption is that Pushkin read Radishchev quite anachronistically, for in the eighteenth century juxtaposing Lomonosov with Bacon could only be seen as high praise. On the other hand, Pushkin may not have misconstrued Radishchev's assessment; rather, he might have been attempting to salvage the comparison with Bacon when it was no longer a laudatory linkage. Pushkin's riposte to Radishchev that "Lomonosov was a great man. Between Peter I and Catherine II he stands alone as a pioneer champion of enlightenment (*prosveshchenie*). He founded our first university. Or rather, he was himself our first university,"[86] seems not incompatible with Radishchev's evaluation. It would be rare indeed to find a study of Lomonosov that does not repeat the above lines from Pushkin.

While he reasoned that Lomonosov's verses—which incidentally he dismissed as having a "harmful" effect on Russian literature—displayed "the absence of any national characteristics and of any originality,"[87] this judgment was not lodged against Lomonosov's science: Pushkin argued that it was essential to keep in mind that "Lomonosov did not value his own poetry and was far

---

[85]   Yeo, "Images of Newton," 278. John Gascoigne regards the relative neglect of Joseph Banks's role as a "statesman of science" as emanating from just such conceptions (Gascoigne, "The Scientist as Patron and Patriotic Symbol," 243).

[86]   Wolff, *Pushkin*, 346; Pushkin, *PSS*, vol. 11, 249. Tatiana Wolff translated *prosveshchenie* as education. From my reading of Pushkin, enlightenment seems to be a better choice. The decision is not merely semantic, however, and depends to an extent on whether or not one accedes to the idea of an eighteenth-century Russian Enlightenment.

[87]   Wolff, *Pushkin*, 347; Pushkin, *PSS*, vol. 11, 249.

more concerned about his chemical experiments." Here he returns to Lomonosov's acclaimed lament to Shuvalov that he was primarily a scientist who, unfortunately for the betterment of the sciences in Russia, was often kept from his laboratory by the onerous duties placed upon him.

A reading of Lomonosov's correspondence, Pushkin notes, quickly reveals his true avocation, for "with what feeling he speaks of science, of education!" Pushkin reprinted a report (*otchet*) that Lomonosov dispatched to the Academy of Sciences detailing his duties in different fields from 1751-56.[88] Academy members commonly sent such accountings to the administration. What is particularly important in this report is that while for each of these years Lomonosov recorded his activities in the study of Russian history, as well as in language and literature, those activities pale in significance beside the summaries of his labors in chemistry and physics which head each year's listing. Lomonosov strongly accentuated the point that he was conducting myriad chemical and physical experiments. His biographers would naturally consider such a stress by Lomonosov on his observational skills a precious source. Pushkin drew the self-evident conclusion from the yearly conspectus that "nothing can give a better understanding of Lomonosov."[89]

Lomonosov's missive to the Academy, routine though it may have been, was utilized by Pushkin to lay stress on Lomonosov's efforts to continually reaffirm his status both at the Academy of Sciences and in Russian society. Being a poet in Pushkin's time was no more secure than being a scientist, or a littérateur, in Lomonosov's. In fact, due to the increasing obsolescence of the older patronage structures, the search for honor and respectability was perhaps

---

88   Ibid., 249-53. Pushkin mistakenly stated that the report was meant for Shuvalov; it was instead addressed to the president of the Academy of Sciences, who at that time (1757) was K. G. Razumovskii. The digest covered Lomonosov's activities from 1751-56, not to 1757 as noted by Pushkin. It was first printed in *Moskovskii telegraf*, part 18, no. 2 (1827): 109-17. See also Lomonosov, *PSS*, vol, 10, 388-93, 783-86.

89   Pushkin, *PSS*, vol. 11, 249.

more problematic. For as Pushkin asked with despair, is it better to be at the mercy of "some rogue or liar" (meaning the rabble), than to write, as Lomonosov did, for "a kind and wise lord"?[90]

Pushkin's allusion was to Shuvalov. If a more personalized patronage mechanism was in place, even in the overly idealized form of an enlightened sponsor, the fashioning of a significant role for oneself in society could be accomplished without relying on the vagaries of public opinion. Pushkin underscored this by citing from a heated letter that Lomonosov sent to Shuvalov in 1761. His patron had attempted secretly to effect a reconciliation between Lomonosov and Sumarokov; Lomonosov, after storming from the scene of the meeting, bitterly complained to Shuvalov that: "I will not play the fool before either illustrious nobles, or for the Lord God Himself."[91] Lomonosov's struggles to maintain what he perceived as his honor much impressed Pushkin.[92] He paraphrased this same passage from Lomonosov in a diary entry of 1834,[93] at a time when he was consumed by slights, both ostensible and real, from the censors, and hence from Nicholas I.

Radishchev assailed Lomonosov's supplications before his Maecenases. Pushkin, exploiting the perspective that historical distance afforded him, brought greater nuance to his verdict. Though Pushkin's own circumstances deeply affected his analysis of Lomonosov's self-presentation, they also seem to have afforded

---

[90]   Wolff, *Pushkin*, 348; Pushkin, *PSS*, vol. 11, 255.

[91]   Ibid., 254. Pushkin was faithful to Lomonosov's wording, which can be found in Lomonosov, *PSS*, vol. 10, 546 (the letter is dated 19 January 1761). Mikhail Pogodin, who published it in *Uraniia. Karmannaia knizhka na 1826 god dlia liubitel'nits i liubitelei russkoi slovesnosti* (Moscow, 1826; reprint, Moscow, 1998), 39-40, first alerted Pushkin to the letter. For more on what led Lomonosov to remonstrate with Shuvalov, see Zhivov, "Pervye russkie literaturnye biografii," 48-49.

[92]   The following mention Pushkin's use of the letter in the context of his uncertain professional situation at the time: David M. Bethea, *Realizing Metaphors: Alexander Pushkin and the Life of a Poet* (Madison, WI: University of Wisconsin Press, 1998), 75; Jones, "The Image of the Author," 61-62; Reyfman, *Trediakovsky*, 221-22; Stennik, *Pushkin i russkaia literatura*, 286.

[93]   Pushkin, *PSS*, vol. 12, 329.

him a greater understanding than Radischchev—who had also been dependent on patronage—was able to bring to bear.[94] With respect to Pushkin's evaluation of Lomonosov's science, like Radishchev he focused on portraying Lomonosov as the first Russian scientist, the heroic opener of new vistas for his people. Pushkin's efforts to rehabilitate the association made with Francis Bacon tellingly underline this. His tone was more deferential than Radishchev's, but it was largely tone that made the difference: in his emphases there were no consequential divergences.

Even Pushkin's re-issuance of Lomonosov's Academy service report featuring chemistry and physics was not employed by him to extol Lomonosov's work as an experimenter, but instead to reinforce Lomonosov's already widely diffused self-identification as personifying the advances that the sciences had brought to Russia. Or to restate Pushkin's celebrated line, which Lomonosov would have heartily agreed with: "he was himself our first university." Still, what made Pushkin's work on Lomonosov so compelling, so copied in the literature, and ultimately so determinant of how the iconic image of Lomonosov as the father of Russian science would appear to later generations was not to be found strictly in his words, which though devoted were also circumspect. The extraordinary veneration that Russian culture lavished on Pushkin allowed his pronouncements on Lomonosov, however much they might have echoed preceding thinkers, to take on an aura of profundity.

The Academy of Sciences launched a far-reaching effort in 1865 to formally enshrine the Lomonosov myth. This had been preceded by a less systematized, though also more intellectually engaging, attempt at institutionalizing and at least partially historicizing remembrances of Russia's first scientist at Moscow University in 1855. This is not to say that in the intervening time, post-Pushkin, Lomonosov's story had attracted a dwindling array of devotees.

---

[94] Or as developed by Svetlana Evdokimova: "In his defense of Lomonosov, Pushkin's narrator demonstrates that, in fact, Lomonosov was courageous enough to defend his own dignity, but that Lomonosov's notion of dignity was obviously different from that of Radishchev." Evdokimova, *Pushkin's Historical Imagination*, 94.

Indeed, works large in scale, scholarly in makeup, and passionate in argumentation were produced.[95]

Some of them remain insightful—or at any rate interesting—studies. What they do not appear to have done is to markedly add to or alter the image of Lomonosov as the national symbol of the sciences that has been sketched out before us. But how this imagery was received, or more precisely, in what manner and for what specific purpose Russian historians, scientists, and writers utilized it, are questions that need now be re-invoked. Of course only tentative answers can be suggested, but perhaps by exploring the ceremonial devotions offered to Lomonosov in mid-nineteenth century Russia a rare opportunity to explore the core of the mythology, and its continued power, presents itself.

---

[95]   Two neglected publications from this period are Ksenofont Polevoi's two-volume—wholly romanticized—historical novel on Lomonosov, *Mikhail Vasil'evich Lomonosov* (Moscow, 1836, 1st vol; St. Petersburg, 1887, reprint of 2nd volume); and the Slavophile Konstantin Aksakov's published thesis on Lomonosov's literary and linguistic activities, *Lomonosov v istorii russkoi literatury i russkogo iazyka* (Moscow, 1846; re-issued during the tercentenary year of Lomonosov's birth, 2011). Aksakov's entry is a dense mixture of Slavonic and Russian linguistic history, Russian literary criticism, Russian history, Hegelianism (of a type), and Russian national sentiment, with Lomonosov serving as the pivot. Lystsov (*Lomonosov v russkoi istoriografii 1750-1850-kh*, 70-258), and Solov'ev and Ushakova (*Otrazhenie estestvennonauchnykh trudov Lomonosova v russkoi literature*, 42-56), are useful guides to a wide array of Russian literary and historical sources from the first half of the nineteenth century that in some manner touch on Lomonosov. The political biases that run through them, however, render their interpretations dubious.

Chapter 4

COMMEMORATING RUSSIA'S
"FIRST SCIENTIST"

At a gathering at Moscow University on 12 January 1855, convened to celebrate the hundredth anniversary of the university,[1] the historian, publisher, academic, and enthusiastic upholder of the Russian past Mikhail Pogodin (1800-75),[2] delivered an address honoring Lomonosov's remarkable achievements in furthering the diffusion of knowledge in eighteenth-century Russia. It was not only as "a pioneer of national

---

[1]   For a guidebook to the ceremonies, see *Stoletnii iubilei Imperatorskago Moskovskago universiteta* (Moscow, 1855).

[2]   Pogodin, the son of a freed serf, was a member of the Academy of Sciences at this time; previously he had held the chair in Russian history at Moscow University. Very close to Sergei Uvarov, the long-serving president of the Academy of Sciences and a former minister of education, Pogodin was a staunch supporter of the conservative "Official Nationality" (Orthodoxy, Autocracy, and Nationality) policies formulated by Uvarov and espoused by the regime of Nicholas I. He was also, along with Stepan Shevyrev, the editor of the journal *Moskvitianin*. For more on Pogodin, consult N. I. Pavlenko, *Mikhail Podogin* (Moscow, 2003); Edward C. Thaden, *The Rise of Historicism in Russia* (New York: Peter Lang, 1999), 90-101; and Cynthia H. Whittaker, *The Origins of Modern Russian Education: An Intellectual Biography of Count Sergei Uvarov, 1786-1855* (Dekalb, IL: Northern Illinois University Press, 1984), 102-07. N. P. Barsukov's twenty-two volume biography of Pogodin, *Zhizn' i trudy M. P. Pogodina* (St. Petersburg, 1888-1910), is an abundant source of information on intellectual life in nineteenth-century Russia. As a "biography," its value lies almost exclusively in its enormous documentary base.

"Lomonosov—The Father of Russian Science"
Painting by A. I. Vasil'ev, 1950

learning, a renowned disseminator of education, a natural scientist, chemist, physicist, geographer, metallurgist, historian, philologist, a writer of prose, and poet,"[3] that he thought Lomonosov should be acclaimed. Perhaps even more meaningful were his lifelong struggles against both his own personal detractors and the wider forces opposed to the development of the sciences in Russia. These antagonists were assuredly one and the same. In battling them, Lomonosov towered above all others in the "intellectual spheres" in which he labored, and "on his mighty shoulders alone carried forward the reforming work of Peter the Great."

To Pogodin, the many tasks Peter the Great took upon himself were each essentially aimed at modernizing Russia. This was a goal that was shared, of course, by Lomonosov. Pogodin was a tenacious defender, and interpreter, of the spirit of the Petrine reforms, and commensurately strove zealously to extol the reputation of Peter the Great.[4] His association of Lomonosov's labors with those of the revered tsar not only underscores Pogodin's abiding regard for Lomonosov's contributions[5] but also reveals the approach he would take to the mythology that surrounded him. For it was not merely

3   M. P. Pogodin, "Vospominanie o Lomonosove," *Moskvitianin*, no. 2 (1855): 1. Pogodin's discourse was apparently not altered for publication (in *Moskvitianin* its length is sixteen pages). "Vospominanie o Lomonosove" is excerpted in *Stoletnii iubilei Imp. Moskovskago universiteta*, 83-94.

4   Pogodin's veneration of Peter the Great as the ruler who brought civilization, defined as much in political, diplomatic, and military terms as in manners, to Russia, is perhaps best confirmed in his essay, "Petr Velikii," in M. P. Pogodin, *Istoriko-kriticheskie otryvki* (Moscow, 1846), 333-63. Pogodin quite appropriately concludes his paean to Peter by quoting from Lomonosov's "Slovo Pokhval'noe blazhennyia pamiati Gosudariu Imperatoru Petru Velikomu" (1755).

5   Linking Lomonosov's biography with Peter the Great's image had become a trope in writings on Lomonosov by the middle of the nineteenth century. During the 1830s and 1840s Belinskii had been especially active in advancing this device. "Lomonosov was the Peter the Great of our literature" (Belinskii, *PSS*, vol. 9 [Moscow, 1955], 674), was one of his much coined phrases. For an explication by Russia's then-foremost critic on why "between Lomonosov and Peter there was a great likeness," please see ibid., vol. 2, 186. Reyfman considers the use of Petrine analogies in literary memoirs in her *Trediakovsky*, 124-25.

his existence as a relic of earlier Russian grandeur that persuaded Pogodin of the persistent value of revisiting his biography. Rather, Pogodin wished to focus on Lomonosov as a living force, with definite, still-existing, accomplishments that could be upheld as his living monument.

"Lomonosov belonged to all of Russia, to the entire Fatherland"—that was of course true and obvious to all—"but he especially belonged to the Petersburg Academy of Sciences, to which he was so devoted, and to Moscow University, in the founding of which he had played a determinant part."[6] As Pogodin tried to persuade his listeners in his lecture, both institutions owed their fundamental character to Lomonosov, so Pogodin spoke of him as "the renowned Russian Academician and Russian Professor."

The Academy of Sciences, and to a lesser degree the Russian university system, began to come under increasing scrutiny from elements within the intelligentsia in the 1850s for their seeming irrelevance, or more gravely their indifference, to the pressing economic, social, and political problems besetting the country.[7]

---

[6]   Pogodin, "Vospominanie o Lomonosove," 2.

[7]   Because of their visible teaching functions, the universities seem to have been seen by critically minded segments of the educated population (the intelligentsia) as less peripheral to everyday concerns than the Academy. The universities, however, had begun to undertake, albeit on a small scale, sustained scientific research work in the 1830s and 1840s. Sergei Uvarov was the initiator of this and many other relatively beneficial reforms in Russian education. Despite Uvarov's evident innovations, he has, owing to his service in Nicholas I's government, regularly been classified as a reactionary. An interesting revisionist account, which persuasively challenges much of the accepted wisdom on Uvarov, is Whittaker's *The Origins of Modern Russian Education* (for Uvarov and science education, see pp. 168-72). See also M. F. Khartanovich, *Uchenoe soslovie Rossii: Imperatorskoi Akademii nauk vtoroi chetverti XIX v.* (St. Petersburg, 1999), which offers a meticulous, and for the most part laudatory, examination of Uvarov's management of the Academy of Sciences. By the 1870s some universities (Moscow and St. Petersburg in particular) had become formidable competitors to the Academy's previous dominance over scientific research in Russia. This evolution of the role of the universities in Russian science was hardly linear, and was dogged by government pedagogic interference, or, worse, financial neglect; nonetheless, the nineteenth century did witness

Consider that in 1855 Russia faced not only defeat in the Crimean War, but with the death of Nicholas I potential political instability. With its large contingent of "foreign scholars,"[8] along with what was perceived as its esoteric commitment to basic research over applied research,[9] the Academy of Sciences encountered especially sharp criticism. Pogodin may have been rigidly statist in his political views, but he was an avowed proponent of educational reforms. As a member of the Academy of Sciences, as well as having been long affiliated with Moscow University, he unequivocally defended both of these bodies as vital to the development of Russia. The

---

a substantive expansion in the base of Russian science beyond the Academy of Sciences to universities as well as to technical institutes. Of the sciences, chemistry, which required the expensive outfitting of laboratories, was perhaps most dependent on government largesse. For a survey of the place of chemistry in Russian universities in the nineteenth century, see Brooks, "Formation of a Community of Chemists in Russia," 147-441 (he attends mainly to chemistry at Kazan', Moscow, and St. Petersburg Universities); and idem, "The Evolution of Chemistry in Russia During the Eighteenth and Nineteenth Centuries," in *The Making of the Chemist: The Social History of Chemistry in Europe, 1789-1914*, ed. David Knight and Helge Kragh (Cambridge: Cambridge University Press, 1998), 163-76.

[8]   Although nationalist-minded historians were prone to exaggeration, it does appear that by the middle of the nineteenth century the Academy of Sciences was widely regarded as a "German institution." Vucinich, *Empire of Knowledge*, 43-56; E. V. Soboleva, *Bor'ba za reorganizatsiiu Peterburgskoi Akademii nauk v seredine XIX veka* (Leningrad, 1971), 43. When the Academy of Sciences rejected Dmitrii Mendeleev's candidacy for membership in 1880, it reignited arguments over the ostensible over-representation of foreigners at the Academy. A veritable firestorm of condemnation greeted the Academy's decision. Many of the press attacks claimed, quite erroneously, that it was the "German" (or otherwise "foreign") members of the Academy who, in their disdain of native Russians and Slavs, had blocked Mendeleev's rightful ascension to their ranks. Analysis of this much-discussed topic is provided by Michael D. Gordin, *A Well-Ordered Thing: Dmitrii Mendeleev and the Shadow of the Periodic Table* (New York: Basic Books, 2005), chapter 5.

[9]   That a distinction between basic and applied research is utterly artificial has in no sense reduced its use in debates over science. The unity of theory and practice is central to representations of Lomonosov. Therefore, when scholars were defending the Academy's position, references to Lomonosov's labors in helping to develop Russia were, as will be shown in Pogodin's memoir, constantly brought forth.

desirability of associating the prestige of the Academy of Sciences and Moscow University with that of a national symbol such as Lomonosov is hence quite apparent.

With respect to the Academy of Sciences, the image of its indelible link with Lomonosov is a conspicuous element in his own self-presentations to Shuvalov. During the eighteenth century the place of the Academy in Russia was all but the reverse from what it would be in Pogodin's time, and Lomonosov had strenuously attempted to juxtapose his own uncertain status with that of the more firmly established Academy of Sciences.[10] Subsequent representations of his scientific work never failed to conflate it with the early years and seemingly lofty repute of that institution.[11] But

---

10   Analogously, Lomonosov's election to honorary membership in the Swedish and Bologna Academies, while it hardly substituted for recognition of him by the more prestigious scientific societies in Paris, Berlin, and London, was at least perceived by Lomonosov as useful in his striving for status in Russia, and was heralded by both him and his biographers.

11   Reputation is a problematic concept, not easily deciphered. It might be comfortably asserted, however, that with the presence of natural philosophers of the caliber of Jakob Hermann, Georg Bilfinger, Georg Krafft, and especially Daniel Bernoulli and Euler, the Academy's first years were a golden age of sorts. By 1741, however, with the departure of Euler — the other scholars mentioned had all ended their service prior to this date — the Academy of Sciences had lost its leading members. Until Euler's return in 1766, Aepinus, and before him Richmann, were perhaps its best-known scientists. Lomonosov was its most prominent Russian natural philosopher, and was, of course, for several years beginning in 1745 its only "Russian scientist" (Stepan Krasheninnikov, appointed professor of botany and natural history in 1750, was the second Russian naturalist to achieve full membership). Even though the Academy's international standing had fallen precipitously since its first decade or so, it still represented within Russia an established institution, while the position of a physicist or chemist, particularly if they had little recognition outside of the country, was barely acknowledged. The St. Petersburg Academy of Sciences has rarely been accorded more that an infrequent aside in surveys of early modern western science. An exception to this is James E. McClellan III, *Science Reorganized: Scientific Societies in the Eighteenth Century* (New York: Columbia University Press, 1985). Despite occasional lapses, most glaringly his failure to firmly situate the Academy in Russian cultural life, McClellan's work does convey its early successes and gives due credit to the high regard in which the Academy was held, for a time, by eighteenth-century European scientists.

as the mythology around Lomonosov's name assumed ever-greater dimensions, his reputation eclipsed that of the Academy. By the middle of the nineteenth century, arguments for the Academy's relevance were increasingly couched in references to its illustrious past.[12] Representations of Lomonosov's devotion to and productivity at the Academy were an indispensable resource in these discussions.

Even given Lomonosov's extensive involvement in the establishment of Moscow University, his portrayal as the sole moving force behind it was quite a bit slower in developing. Indeed, none of his eighteenth-century biographers saw fit to mention Lomonosov's "founding" of Moscow University, or even his association with it, in their studies. Severgin also ignored the connection in his later evaluation, and when the fiftieth anniversary of the university was observed on 30 June 1805 Lomonosov's role was not acknowledged in P. A. Sokhatskii's expansive oration surveying the university's history.[13] The principal accolades were bestowed on Ivan Shuvalov

---

See also Iu. Kh. Kopelevich's fine *Vozniknovenie nauchnykh akademii: seredina XVII – seredina XVIII v.* (Leningrad, 1974), 176-229. Here and in her later monograph, *Osnovanie Peterburgskoi Akademii nauk*, Kopelevich investigates the founding of the Academy of Sciences, its initial relationships, institutional and intellectual, with leading European natural philosophers and with other scientific academies, and the considerable esteem in which she maintains it was held.

[12] Pekarskii's two-volume history of the Academy of Sciences, *Istoriia Akademii nauk* (1870-73), represents an apotheosis of the trend. Biographically structured, Pekarskii presents the activities at the Academy of Sciences of its members over approximately the first half-century of its existence. His memoirs of, among others, Euler, Bernoulli, Richmann, Adodurov (the first Russian adjunct, his work was, at least ostensibly, in "higher mathematics"), and Lomonosov (at approximately 700 pages, his entry composes most of the second volume) contain a wealth of primary and secondary source materials and are still quite indispensable in the study of their subjects. Pekarskii explicitly connected the Academy—as it stood in the last decades of the nineteenth century—to the glory of its early years.

[13] P. A. Sokhatskii, *Slovo na poluvekovoi iubilei Moskovskogo universiteta* (Moscow, 1805). See also Shevyrev's rather skeletal account of the day's proceedings in *Istoriia Imp. Moskovskogo universiteta*, 358-61. The day was marked by the usual fare of jubilees: church services, speeches (six on this occastion), and presentations by dignitaries representing the state, the church, and the university. In the evening, a display of illuminations

and the Empress Elizabeth. Lavrentii Blumentrost (who served briefly as the university's co-curator with Shuvalov) also received acclaim. A complete turnaround occurred in the following decades, with Lomonosov's name eclipsing all others. Pushkin's oft-quoted remark about Lomonosov that "he founded our first university" nicely summarizes the evolving mythology. By the time of the university's centennial, Russian's "first" university was increasingly viewed as essentially Lomonosov's creation.[14]

Despite Pogodin's assertion that "we can all recite by memory the life of Lomonosov,"[15] he was not dissuaded from recalling for his audience, at length, the long-established details that constituted Lomonosov's biography. If done in the proper fashion, conveying the heroic qualities of Lomonosov was highly instructive, for his life itself, which was punctuated throughout by incessant struggles engaged in and challenges overcome, "is all together a miraculous picture, in fact it is one of the most striking of such images in our history, which is filled with abundant miracles." He clearly set

---

closed the ceremonies. Although Mikhail Murav'ev was in 1805 the trustee of Moscow University, he offered no speech, or other work, that dealt with Lomonosov's involvement with the university.

[14] Russia's first university was not Moscow University; rather it was a small institution, nominally termed a university, which had been founded as part of the Academy of Sciences (1725) that earned that designation. With some difficulty it enrolled its first students, numbering eight, in 1726. All of them were imported from abroad. Lomonosov directed the Academy's educational efforts, which included a gymnasium, for several years. It would seem that he was unsuccessful in reversing the university's decline (the gymnasium was more successful). The Academy was never able to attract either a sufficient amount of students to the university or to hold on to those that they did enroll. The plight of the academic university is ably presented in Ludmilla Schulze, "The Russification of the St. Petersburg Academy of Sciences and Arts in the Eighteenth Century," *British Journal for the History of Science* 18 (November 1985): 305-35; Galina Smagina, *Akademiia nauk i rossiiskaia shkola. Vtoraia polovina XVIII veka* (St. Petersburg, 1996); and D. A. Tolstoi, *Akademicheskii universitet v XVIII stoletii po rukopisnym dokumentam Arkhiva Akademii nauk* (St. Petersburg, 1885). For a brief account of the university's disappearance, which occurred without fanfare in the late 1760s, see Ostrovitianov, *Istoriia Akademii nauk*, vol. 1, 420-23.

[15] Pogodin, "Vospominanie o Lomonosove," 3.

before the audience, both those assembled and those in posterity, the idea, and even more the obligation, of emulating it—for the ultimate purpose of biographically returning again and again to idealized figures, scientific, literary, political, and military or any other constructed type, is mainly educative.

In the pages of his journal *Moskvitianin*, Pogodin was the first to publish Staehlin's "Traits and Anecdotes for a Biography of Lomonosov."[16] His knowledge of it, as well as of Verevkin's and Novikov's short biographies, was thorough. Moreover, to an extent Pogodin's lecture is a restatement of these foundational texts. He not only singles out many of the same details and anecdotes—the latter taken from Staehlin—but his strong focus on Lomonosov's intrepid youth duplicates their belief in the necessary precursive attributes of greatness. Pogodin does, however, add a sharper polemical thrust, directed most clearly against the enemies of learning, or rather the enemies of the Petrine/Lomonosov heritage, than was present in his predecessor's memoirs. Additionally, while Lomonosov's earlier biographers employed his mythology quite broadly for the satisfaction of contemporary agendas—an essential factor explaining the persistence of myth[17]—Pogodin wielded it for

---

16   *Moskvitianin*, no. 1 (1850): 1-14. Pogodin disbelieved Staehlin's authorship, and was convinced that Damaskin (the one-time rector of the Slavo-Greco-Latin Academy) was the actual memoirist (see footnote 35 in Chapter 2). This may have resulted from Pogodin's occasional willingness to deprecate German influence at the Academy, both in his own time and during Lomonosov's. In *Moskvitianin*, no. 3 (1853): 22-25, Pogodin also published Staehlin's "Konspekt pokhval'nogo slova Lomonosovu." For more on Pogodin's interest in the fate, and recovery, of Lomonosov's writings, see Kuliabko and Beshenkovskii, *Sud'ba biblioteki i arkhiva Lomonosova*, 27, 85-98; Zubov, *Istoriografiia*, 348.

17   Roland Barthes's view of myth is far from sanguine; indeed, his vision of it is wholly oppressive. "It is a kind of ideal servant: it prepares all things, brings them, lays them out, the master arrives, it silently disappears: all that is left for one to do is to enjoy this beautiful object without wondering where it comes from…. Nothing is produced, nothing is chosen: all one has to do is to possess these new objects from which all soiling trace of origin has been removed. This miraculous evaporation of history is another form of a concept common to bourgeois myths: the irresponsibility of man." Roland Barthes, *Mythologies*, trans. Annette Lavers (New York, 1972), 151.

very specific cultural and educational ends. This was an indication
of the myth's maturation.

Astonishing to Pogodin was the fact that the responsibility of
"planting European science into Russian soil would be entrusted
by fate to a simple peasant, who was born in a poor peasant's
hut."[18] Lomonosov's origins in a distant village near the White
Sea continued to have great resonance in the scholarship, as it still
does,[19] and Pogodin seamlessly wove his own marvel at it into his
speech. Especially consequential were Lomonosov's early successes
in overturning all obstacles on his journey to enlightenment. Here
his use of Staehlin and Verevkin is perceptible. Pogodin's own
unassuming economic and social background (his family had
been emancipated from serfdom when he was a child) confers an
unmistakable impression of sincerity to his emotive description of
Lomonosov's ascent to the heights of scientific and literary success.
It also harmonized with the at best vaguely-articulated Petrine
objective of a selective meritocracy.

Lomonosov's passion for learning received great praise from
the speaker. What perhaps most concretely defined his curious
mind during his youth, as well as framing his later interests, was his
reputed ceaseless perusal, and eventual possession, of Smotritskii's
Grammar and the *Arithmetic* of Magnitskii.[20] That this began place
in his early adolescence had certain import. With an unspoken nod
towards Verevkin, Pogodin referred to these texts as Lomonosov's
"gateways to learning." Novikov had listed Polotskii's *Psalter*
as meaningful in Lomonosov's education, and Pogodin cited its
influence on Lomonosov with the utmost respect. These writings,
each profoundly significant in Russian history, not only indicate
Lomonosov's "hunger for knowledge," exceptional as it was, but
also implicitly direct the listener to the deeper impact that Peter's
cultural reforms had on Russia. Along with urging the spread
of education, the preservation of the Russian past was of great

---

18   Pogodin, "Vospominanie o Lomonosove," 3.

19   See Lebedev, *Lomonosov*, 16-30.

20   Pogodin, "Vospominanie o Lomonosove," 4.

importance to Pogodin. Those aspects of Lomonosov's life that carried the clearest corollaries came in for extended consideration.

Lomonosov's preternatural successes at the Slavo-Greco-Latin Academy, where his quest for an education in the sciences could never be fully completed, represents much more than Lomonosov's own individual strivings. For although Peter had brought science to Russia, his successors had failed to encourage its growth properly.[21] Lomonosov, by contrast, who was perhaps the finest progeny of Peter's transformations, had by his own work at the Academy of Sciences and through his inaugurating of Moscow University returned Russia to the path that Peter had marked out of it. Pogodin forcefully imparted this notion to his listeners, for it provided an apparently clear pattern of historical continuity from Peter the Great's age, through Lomonosov's, to his own.

Lomonosov's biography as first fashioned in his letters to Shuvalov and later given shape by eighteenth-century memoirists was still almost completely structured by questions of character. Selfless labor and a near superhuman productivity were what determined the cultural meaning of his work. His genius was of course extolled, but it was the moral qualities he exhibited, including a fiery resolve to advance the sciences in Russia, which inspired fascination with his life. What his papers might have revealed of his chemistry and physics was not dismissed, but the minutiae were

---

21   Many of Lomonosov's compositions are permeated with this theme. Wonderfully illustrative of this is his plea—which became a cliché in the literature—to the Empress Elizabeth, and indeed to Russians, "to show with zeal, that the Russian land can give birth to its own Platos and quick-witted Newtons." Lomonosov, *PSS*, vol. 8, 206 (from his "Oda na den' vosshestviia na Vserossiiskii prestol Eia Velichestva Gosudaryni Imperatritsy Elisavety Petrovny," 1747). Lomonosov's various odes and oratorical prose dedicated to the Empress Elizabeth, while expectedly full of praise for her achievements, quite nearly without exception contain a distinct subtext. As the daughter of Peter the Great, and his successor in body and spirit, she must continue with the tasks he set before subsequent generations, tasks that were lost sight of during the reigns of his more immediate successors, Peter II and Anna. Lomonosov's eloquent calls to the empress were also a clear example of his own search for status, for after all, was he himself not one of the descendants of Peter the Great, a Russian Plato or Newton?

unnecessary, even cumbersome, to wide-ranging efforts to utilize the mythology as a universal model for subsequent generations of Russians.[22]

Perhaps the archetypical nineteenth-century illustration of a didactic scientific biography is David Brewster's two-volume study of Newton. The work, issued the same year as Pogodin's assessment of Lomonosov, was premised on his belief that "if we look for instruction from the opinions of ordinary men, and watch their conduct as an exemplar of our own, how interesting must it be to follow the most exalted genius through the labyrinth of common life."[23] For Brewster, Newton's life embodied the ideal

---

[22]   Much the better if they were younger generations of Russian scientists. Geoffrey Cantor's study of biographies devoted to Michael Faraday offers an instructive analogy. An outpouring of works on Faraday appeared in a short space of time following his death in 1867, and they fulfilled a variety of functions: "for some authors, he became the great discoverer of nature's secrets, while for others he was the Christian philosopher *par excellence*, or the leading public lecturer, or the scientist with refined sensibilities— to mention but a few." The value of these memoirs was not limited to the edification of youth, for a "subtext of such narratives is that readers— especially prospective scientists—should adopt Faraday's methods and attitudes as their model." Cantor, "Public Images of Michael Faraday," 172. The absence of women, or rather "unacceptable" women, as role models in scientific biography is explored in Martha Vicinus, "'Tactful Organising and Executive Power': Biographies of Florence Nightingale for Girls," in Shortland and Yeo, *Telling Lives in Science*, 195-213.

[23]   David Brewster, *Memoirs of the Life, Writings, and Discoveries of Sir Isaac Newton*, vol. 1 (Edinburgh: Thomas Constable, 1855), 3. In a narrow sense, Brewster's work is directed at Francis Baily's *An Account of the Revd. John Flamsteed, The First Astronomer-Royal* (1835). Baily's life of Flamsteed, which, utilizing a vast array of Flamsteed's correspondence, attempted to reinterpret the dispute between the royal astronomer and Newton at the latter's expense, had caused no end of upset among Newton's devotees. Brewster hoped to rectify Baily's assault on Newton's reputation, commenting pointedly in his preface that: "I trust that I have been able, though at a greater length than I could have wished, to defend the illustrious subject of this work against a system of calumny and misrepresentation unexampled in the history of science" (*Memoirs of Newton*, vol. 1, XI). Brewster, Newton historiography, and the development of scientific memoirs are discussed in Hall, *Isaac Newton*, 181-85; Higgitt, *Recreating Newton*, 43-68, 129-57; Adrian Rice, "Augustus De Morgan: Historian of Science," *History of Science* 34,

synthesis of intellect and integrity. In his still consequential study, Newton's science was, in contrast to then-extant biographies of Lomonosov, given sustained analysis. But conceptions of Newton's supremely elevated character were inextricably bound up with his genius. As averred by the author, for the curious student:

> The writings and the life of Sir Isaac Newton abound with the richest counsel. Here the philosopher will learn the art of patient observation by which alone he can acquire an immortal name; the moralist will trace the lineaments of a character exhibiting all the symmetry of which our imperfect nature is susceptible; and the Christian will contemplate with delight the High Priest of Science quitting the study of the material universe, the scene of his intellectual triumphs, to investigate with humility and reverence the mysteries of his faith.[24]

Given the saintly quality associated with most eighteenth- and nineteenth-century views of Newton, to mirror one's life on his was, for lesser souls, a nearly impossible goal to attain. This, however, was a distinguishing characteristic of such biographies; it is the heroic ideal that made them such attractive stories. As written by Brewster, the "life" of Newton was identical with his science. Unyielding perceptions of Lomonosov's pioneering deeds still obscured, or impeded, a careful examination of his actual work; an attempted equivalence between the two was not yet witnessed in the literature.

In evaluating Lomonosov's science, Pogodin turned to familiar tales. Lomonosov's journeying to Marburg "to attend lectures with the celebrated philosopher and mathematician of the day, and a student of Leibniz, Wolff"[25] remained an engrossing episode, for it was during this sojourn abroad, which Pogodin narrated quite

---

no. 104 (June 1996): 212-19; Paul Theerman, "Unaccustomed Role: The Scientist as Historical Biographer — Two Nineteenth-Century Portrayals of Newton," *Biography* 8, no. 2 (Spring 1985): 145-62; and Yeo, "Images of Newton," 270-79.

24   Brewster, *Memoirs of Newton*, vol. 1, 3.

25   Pogodin, "Vospominanie o Lomonosove," 7.

fully, that this provincial fisherman's son had "opened up before his inquisitive mind a new world." By virtue of his "talents, work ethic, and enterprising nature, he learned all that it was possible to acquire, he mastered modern science," and of inestimably greater portent, he discovered that he could himself teach his countrymen to follow the proper path, that he could "introduce [or reintroduce] science into Russia, into his beloved fatherland." That which Pogodin termed "European science" received steadfast deference. Russian judgments of Christian Wolff's legacy had become quite critical by this time,[26] but Pogodin's review, Lomonosov's stellar scientific education in the Germanies, the specifics of which he did not share with his audience, and his tutelage under such an illustrious teacher, remained notable.

After Lomonosov's return to St. Petersburg, and following his brief service as an adjunct in physics at the Academy, he was named to the chair of chemistry in 1745 (due to his reliance on the chronology supplied by Staehlin and Verevkin, Pogodin mistakenly dates it as 1746). With this begins the period of Lomonosov's lasting achievements.[27] While Pogodin was convinced of Lomonosov's gifts as a chemist and physicist, it was not Lomonosov's corpuscular/mechanical investigations that drew his historical interest. Instead, his praise was riveted on Lomonosov's educational and organizing activities at the Academy. By accomplishing such "firsts" as

---

[26]   Sukhomlinov's "Lomonosov—student Marburgskogo universiteta" (1861), is a sound account of Lomonosov's schooling in Germany. That said, it is better on Lomonosov's literary and linguistic training in Marburg than on his education in the natural sciences and mathematics. By failing to more thoroughly consider either Wolff's metaphysics or his treatment of mathematics, Sukhomlinov's essay typifies many of the difficulties in attempting to situate Wolff in studies of Lomonosov. He essentially argues that Lomonosov received a solid grounding in the sciences in Germany, including an unfortunate overexposure to Wolff's metaphysical leanings, but then—and here Sukhomlinov offers no demonstrable proofs—promptly jettisoned these conceptions when commencing his own work at the Academy of Sciences. Thus, Lomonosov took the best from Wolff, while ignoring the more problematic monadology of his mentor.

[27]   For the author's summary of Lomonosov's more public feats as an academician, see Pogodin, "Vospominanie o Lomonosove," 9-14.

delivering widely attended public lectures on chemistry and physics, as well as setting up the Academy's chemical laboratory, he laid the foundations for chemistry in Russia. The information discovered through his experiments and the research he conducted as an academician were disseminated in his writings,[28] and were therefore implicitly influential.

Of evidently lasting consequence, and not surprisingly esteemed by Pogodin, was Lomonosov's direction of the gymnasium and university attached to the Academy. Dispirited by the poor functioning of the Academy's educational mission, Lomonosov designed several projects for its restructuring.[29] He also sought to reform the Academy of Sciences's internal administration. Involvement of a supervisory nature with the Academy's publishing

---

[28]   Among the purely "scientific" papers referred to (ibid., 10) were those that earlier scholars either examined or alluded to repeatedly, including: *Oration on the Usefulness of Chemistry*; *Letter on the Usefulness of Glass* (a discourse that Pogodin connected, accurately, to Lomonosov's work on mosaics); *Discourse about Air Phenomena, Caused by Electricity*; *Oration on the Origins of Light, Representing a New Theory of Colors*; *A Discourse on the Birth of Metals from the Quaking of the Earth*; *Discourse on Greater Accuracy of the Sea Route*; *The Appearance of Venus Before the Sun*; and *First Principles of Metallurgy or Mining*. Also mentioned were some of Lomonosov's literary and linguistic writings, such as the *Russian Grammar*, *Rhetoric*, and select panegyric speeches. Pogodin was not merely impressed by the content of this "splendid list," what most struck him was the "amazing speed with which he completed one paper after the other."

[29]   Many official documents related to Lomonosov's interest in the Academy's educating mandate are found in Lomonosov, *PSS*, vol. 9, 435-611, 847-933. Little creed should be given to notions of the supposed aversion of Russians to organized secular learning in the eighteenth century (see this argument throughout Vucinich, *Science in Russian Culture: A History to 1860*). A perusal of the above educational records buttresses the position that a severe lack of governmental support resulted in pitiable university and gymnasium facilities. The conditions that the students faced, materially and in the quality of instructors, was not enviable. This is perhaps more the reason why so many potential students were averse to attending the Academy's schools than the baleful influence of Russian Orthodoxy was. Although given Lomonosov's tendencies to engage in academic battles over resources and privileges, along with the commensurately rhetorical nature of his pleading, these documents have to be read with care, they are nonetheless informative.

endeavors, the planning of scientific expeditions, the composition of odes, and the production of fireworks—specifically inscriptions for them: each of these activities received the speaker's attention.

Such services also had a conspicuous public character, for even Lomonosov's work in the laboratory was to be commended mainly for its representative value: this is what Russians were capable of doing. Pogodin's evaluation does not differ from those of previous generations of biographers in a further respect; he is most concerned with the practical advantages of Lomonosov's contributions to Russian culture. He simply secured the imagery more clearly to an institutional setting. A demonstration that important research of a decidedly pragmatic nature had been successfully pursued within the Academy would prove quite useful in debates over its contemporary significance.[30]

---

[30] But as David Joravsky queried, the issue of the pragmatism of science in Russia "points to a cursed question, as the Russians say, which has dogged Russian science through all its periods of sharply fluctuating fortunes into its disastrous present: What *is* practical for the pursuit of science in a backward province or country?" David Joravsky, "The Perpetual Province: 'Ever Climbing up the Climbing Wave'," *Russian Review* 57, no. 1 (January 1998): 3. What is "practical" is that, of course, which commands the resources at the moment. An apt comparison with the use of Lomonosov in debates over pure and abstract science is with Louis Pasteur's location in French scientific, historical, and political discussions. As noted by Chrisitiane Sinding, "The constant exchange between empirical and scientific knowledge, and scientists' reworking of practical and technical problems and successes, serve to erase the distinction between applied and pure science. But when Pasteur's commemorators—whether scientists, philosophers or historians— allude to the practical and empirical aspects of his work, they just point out that he was brilliantly able to handle the constant exchange and avoid the issue of the origin of empirical knowledge, because it would bring them to the boundaries between science and nonscience, and between scientists and nonscientists." Pasteur's memorializers, especially among scientists, having realized the worth of his name in patronage mechanisms, "do not want to challenge the idea that basic science leads to true knowledge, which in turn gives birth to applied science, which leads to the solution of all human problems." Christiane Sinding, "Claude Bernard and Louis Pasteur: Contrasting Images through Public Commemorations," in Abir-Am and Elliot, *Commemorative Practices in Science*, 85. Not to press the analogy too far, but comparable to the Academy of Sciences's and Moscow University's roles in lauding Lomonosov's reputation is the Pasteur Institute's in honoring its

Lomonosov's electrical experiments, which were ineradicably combined in the historic imagination with the dramatic death of his collaborator Richmann, had lost none of their power to engage the attention of an audience, and Pogodin adroitly interpreted its import. Rather than relying only on his own narrative skills, Pogodin relayed, in full, Lomonosov's letter to Shuvalov in which he vividly described Richmann's death, touched on some of the details of their shared research, and ended with the hope that this tragedy should not have a detrimental effect on the development of science in Russia.[31] The image of Richmann's martyrdom and Lomonosov's subsequent perseverance remained a powerfully inspiring one. Pogodin stressed that Lomonosov's well-publicized investigations, which were "judged by the entire European scientific world,"[32] had brought fame both to him and to the Academy from outside of Russia. It was not only Lomonosov's individual honor that was heralded by Pogodin—and in the mythology Richmann's role, however courageous, was purely in support of Lomonosov—but also Russia's scientific heritage. Marked by valorous deeds, most strikingly Lomonosov's, it deserved the highest regard.

In ascribing to Lomonosov an honored position among non-Russian scientists, he asserted that Lomonosov had "proposed discoveries, which Europe had marveled at when made by Franklin and Rumford [Sir Benjamin Thompson, Count Rumford], and half a century later would be seen in the writings of Arago [Dominique Françoise Arago] and Gumbol'dt [Alexander von Humboldt]."[33]

---

founder. How the "pasteureans" extended the myth of Pasteur over French medical and scientific life is examined in Bruno Latour, *The Pasteurization of France*, trans. Alan Sheridan and John Law (Cambridge, MA: Harvard University Press, 1988).

31   Pogodin, "Vospominanie o Lomonosove," 12-13.

32   Ibid., 12.

33   Ibid., 13. Thompson (1753-1814) worked most productively on heat conduction and ballistics, Arago (1786-1853) on electricity, magnetism, and light, and the encyclopedic Humboldt (1769-1859) on a variety of topics related to physical geography. Pogodin makes a reference, without further elucidation, to the work of D. M. Perevoshchikov, whose juxtapositions of Franklin and Lomonosov were previously cited. Perevoshchikov and Pogodin served

Given the configuration of the prevailing mythology, the significant comparison is with Franklin. Incorporating into the imagery additional celebrated natural philosophers reinforces the notion that either Lomonosov shared the honor of discovery with other, undeservedly more recognized, West European scientists or that his hypotheses anticipated theirs. Although Pogodin makes the perfunctory likening of Lomonosov to Franklin, for the comparison is directed at the honors that Pogodin clearly insinuates that they should have both received. As for the content of the experiments themselves, in the reading given them, except for the association with Franklin and electricity, they are somewhat beside the point.

None of Lomonosov's efforts in either advancing his own work or spurring the growth of science and education was easily accomplished, for distinct from the inherent difficulties of the scientific research itself, he was continually beset by foes within the Academy of Sciences. Central to the mythology, this theme of enmity was not utilized by Pogodin to berate specific antagonists of Lomonosov; instead he invoked it to question the motives of those who would hinder Lomonosov's vital labors on behalf of Russian science and education. Almost as an afterthought, Pogodin did designate as obstructionist the "German element of those days" in the sciences, who, "however worthy they were otherwise," clashed, by their very makeup it would seem, with "the Russian nature."[34] Passionate in his defense of what he believed in, which was most assuredly the cultural "advancement" of Russia, Lomonosov, to Pogodin, personified the patently tempestuous Russian character.

This signaling out of select enemies seems to have been inserted primarily for rhetorical purposes. Lomonosov's incessant struggles against enemies of both himself and of progress in general, as he or rather Pogodin defined it, required opponents. Despite this animosity, however, the speaker reassured all that:

---

together both at Moscow University and at the Academy of Sciences. Some knowledge by Pogodin of Perevoshchikov's writings on Lomonosov, which largely concentrate on his physical experiments, thus might be assumed.

34   Ibid., 15.

Lomonosov did not become dejected, and throughout his entire life he fought against all his adversaries, he argued, complained, entreated, wept, mocked, cursed, justified himself, and besides all of this, he worked, he worked— and he did not look back with any regret on that which was placed into his hands from above. For him science was always above everything else. The diffusion of science throughout the fatherland was dearer to him than anything. The glory of Russia precious above all else.[35]

Ending with Lomonosov's own words, "candidly" expressed, revealing of his "noble hopes and wishes," Pogodin repeated for the assembly Lomonosov's dirge to Staehlin as to whether the sciences would survive in Russia without him. Pogodin's revisiting of Staehlin's famous anecdote, where Lomonosov's love for Russia is so resolutely demonstrated, perfectly encapsulates his assessment of Lomonosov. The Academy of Sciences and Moscow University were the tools by which Lomonosov tried to achieve his goals of a Russia where "European science" was firmly grounded.

Lomonosov's connection with the Academy of Sciences was the subject of far more sustained consideration in Pogodin's speech than were his exertions in setting up Moscow University. That Moscow University was largely the initiative of Lomonosov was by this time almost presumed, and their mutual link was strongly accented by the occasion itself, the jubilee celebrations, during which Pogodin outlined Lomonosov's worth. In subsequent decades Lomonosov's association with the university would grow ever more integral to their respective identities, so much so that Shuvalov's arguably more important efforts in securing its establishment would be severely downgraded.[36]

---

35   Ibid., 16.

36   The nadir for Shuvalov's reputation came during the Soviet period, when as an aristocrat and favorite of the Empress Elizabeth, his biography was ill-suited to the Marxist-Leninist dicta that dominated eighteenth-century historiography. M. T. Beliavskii's *M. V. Lomonosov i osnovanie Moskovskogo universiteta* (Moscow, 1955) was the first significant study reviewing the university's founding and first decades of growth since

When Moscow University celebrated its 185th anniversary in 1940, "the name of its founder M. V. Lomonosov was awarded to it."[37] This decision merely formalized a long process that saw the imagery around Lomonosov's life attain continually more magnificent proportions. If Pushkin was not the first to claim Moscow University for Lomonosov, his avowal has obscured those of other writers. Successive generations of scholars made the university an essential part of his biography. If a myth is to sustain itself, a steady accretion of narrative detail to it is requisite not only to its survival but also to its vibrancy. During the Soviet period the exaltations of Lomonosov's achievements may have reached an officially encouraged pinnacle; the underlying elements that constituted such acclaim, however, were developed far earlier.

Also recognizing Lomonosov for the Moscow University Jubilee commemorations was N. A. Liubimov (1830-97),[38] a professor of physics at the school. His essay, "Lomonosov as a Physicist" ("Lomonosov kak fizik"),[39] is a remarkable anomaly in the

---

Shevyrev's a century earlier, and in marked contrast to Shevyrev, Beliavskii aggressively promotes Lomonosov's role over Shuvalov's. Shuvalov's work is also altogether minimized in comparison with Lomonosov's in *Istoriia Moskovskogo universiteta*, vol. 1 (Moscow, 1955). Beliavskii was one of the principal editors of this jubilee (200th anniversary) collection. It is, however, a considerably less exacting treatment than that afforded the beginnings of the university in his above monograph. For a recent assessment of the university's founding, worthwhile is Kulakova's *Universitetskoe prostranstvo i ego obitateli*, 25-48. Kulakova's account supports Shevryev's balanced exposition underling Shuvalov's seminal contributions, with considerable intellectual support afforded him by Lomonosov, to the university's foundation.

[37]   As announced in a decree of the Presidium of the Supreme Soviet of the USSR (dated 7 May 1940), which is cited in Beliavskii, *Lomonosov i osnovanie Moskovskogo universiteta*, 270.

[38]   For more on Liubimov, please see "Liubimov, Nikolai Alekseevich," in *Entsiklopedicheskii slovar'* (*Brockhaus-Efron*), vol. 18 (St. Petersburg, 1898), 209.

[39]   N. A. Liubimov, "Lomonosov kak fizik," in *V vospominanie 12-go ianvaria 1855 goda. Ucheno-literaturnye stat'i professorov i prepodavatelei Moskovskogo universiteta* (Moscow, 1855), 3-35. Liubimov's composition was issued for the jubilee as part of a diverse collection of research papers by Moscow

literature dedicated to the Russian polymath. While it is a more
extensive assessment of his science than nearly all others issued
previously, it has also largely been dismissed, or perhaps one might
better say ignored, in the historiography. With the noteworthy
exception of Pekarskii's biography of Lomonosov[40] (which does
refer to, if not deeply rely on, Liubimov's account), there has been,
discounting the very occasional citation, practically no examination
of it. While Liubimov's conservative educational views made
him a controversial figure in his day,[41] this may have been less

---

University professors. Liubimov also authored a full biography of
Lomonosov, *Zhizn' i trudy Lomonosova: s prilozheniem ego portreta* (Moscow,
1872). This entry into the Lomonosov industry has been neglected, rather
undeservedly so, for it is a fairly comprehensive, and judicious, introduction
to his life.

[40]   Pekarskii is primarily interested in chronicling Lomonosov's life at the
Academy of Sciences, which he does extremely well. He does not, however,
appraise Lomonosov's scientific skills; instead, he prints or reprints selected
correspondence and excerpts from Lomonosov's more accessible papers,
and allows the reader largely to draw their own conclusions. The implied
judgement is that while Lomonosov was an original thinker for his times, his
ideas, in terms of contemporary science, are of mainly antiquarian curiosity.
Furthermore, not being a scientist, Pekarskii seems to have believed them
to be beyond his purview. He did, at times, when discussing Lomonosov's
physics, either refer the reader to Liubimov's "Lomonosov kak fizik" or
quote from it. Perevoshchikov is cited less substantively. As for chemistry,
Pekarskii begged off a rigorous inspection of Lomonosov's papers, and
indeed stated that: "until this time none of our specialists has taken the
time to examine and evaluate from an historical perspective the meaning
of Lomonosov's work in chemistry, and therefore I have had to be satisfied
with a superficial sketch." Pekarskii, *Istoriia Akademii nauk*, vol. 2, 450-51
(I was first alerted to this in Leicester, *Lomonosov on the Corpuscular Theory*, 41).
N. E. Liaskovskii, a professor of chemistry at Moscow University, provided
Pekarskii's "sketch." See his "Lomonosov kak khimik," *Prazdnovanie stoletnei
godovshchiny Lomonosova 4 aprelia 1765-1865 g. Imperatorskim Moskovskim
universitetom v torzhestvennom sobranii aprelia 11-go dnia* (Moscow, 1865),
57-66. This short essay, which was first delivered at the Lomonosov Jubilee
held at Moscow University in 1865, is entirely eulogistic in tone and in no
sense scrutinizes Lomonosov's chemical dissertations.

[41]   Liubimov's participation on a commission set up in 1875 by the minister
of education, Dmitrii Tolstoi, to revise the comparatively liberal university
statute of 1863 caused consternation among his colleagues at Moscow
University. Many of them publicly condemned this decision and ostracized

a determinant on his exclusion from Lomonosov studies than the fact that his assessment of Lomonosov's science is so starkly at odds with accepted notions of his unassailable talents. Whatever the ultimate reason, the combination of Liubimov's marginality to the intelligentsia and his decidedly non-hagiographic views on Russia's first scientist earmarked him and his work for the periphery.[42]

This volume will omit a discussion of the minutiae of Liubimov's paper, as it is more profitable for our purposes to scrutinize aspects of his evaluation that directly challenged central tenets of the Lomonosov mythology. The title of his article only approximates its content, for he discourses at some length on Lomonosov's presumed knowledge of contemporary physics, mathematics, and chemistry. Liubimov's more provocative critiques of the Lomonosov legend, subsumed under a rather schematic effort to situate him within eighteenth-century debates between what for semantic ease will be termed Cartesians and Newtonians, wrestle ultimately with Lomonosov's place in the history of science. Discussions of the originality of Lomonosov's work, its meaning to later generations of scientists—primarily physicists and chemists—, the influence of Christian Wolff, Lomonosov's abilities as a mathematician, and his

---

him. See M. M. Kovalevskii, "Moskovskii universitet v kontse 70-ikh i nachale 80-ikh godov proshlogo veka (lichnye vospominaniia)," *Vestnik Evropy: zhurnal nauki—politiki—literatury* (May 1910): 185-87. Kovalevskii, a strong proponent of autonomous university governance, was from the late 1870s on the juridical faculty of Moscow University. He was later (1887) dismissed from the university because of his "liberal" views. His assessment of Liubimov is pungent. See also James C. McClelland, *Autocrats and Academics: Education, Culture, and Society in Tsarist Russia* (Chicago: University of Chicago Press, 1979), 65-66; and E. V. Soboleva, *Organizatsiia nauki v poreformennoi Rossii* (Leningrad, 1983), 29-30. Soboleva's polemical effort, in which she censures Liubimov as a "reactionary," is interesting principally as an example of Liubimov's near total absence from more recent scholarship on the history of Russian science and education.

[42]   The failure of the intelligentsia to include politically or ideologically unpalatable figures in their self-generated "genealogies," with the expected debilitating effects on the writing of history in Russia, is persuasively argued in Marc Raeff's "Russian Intellectual History and Its Historiography," *Forschungen Zur Osteuropäischen Geschichte* 25 (1978): 300-01.

responses to Newtonianism, are all examined by Liubimov. He also inspects the persistent analogies between Franklin and Lomonosov.

Initially, Liubimov's treatise hints at another panegyric as he appropriates Lomonosov for the sciences, declaring that:

> Contemporaries knew Lomonosov more as a poet and writer, than as a scientist. For us he is the first Russian scientist. His literary works, however imbued they are with the most profound feelings, in their essence though they are the products of the highest intelligence, they are the not the creations of a genius.... But in his works in the natural sciences the scientific genius of Lomonosov is expressed fully. Here his words are infused with a clear understanding, a strong conviction, and they reveal a pure Russian cast of mind.[43]

As the age of encyclopedic natural philosophers receded ever further, the shaping of Lomonosov's life by later writers continued apace. Liubimov does very briefly touch on his other activities, mainly literary and linguistic, but though these areas are perhaps due peripheral exploration, "physics and chemistry were the favorite subjects of Lomonosov."

A letter from Lomonosov to Shuvalov, which Liubimov excerpted in full, persuasively accents this proposition. Having been asked by Shuvalov to commence with his writings on Russian history, and presumably "to abandon his physical and chemical investigations,"[44] Lomonosov demurred, arguing that his scientific work "can bring no less benefit to my native land, than my first occupation." To students of Lomonosov's science his "first occupation" was as a chemist and physicist, and as had Severgin and Pushkin, Liubimov used this letter to fortify Lomonosov's standing as a scientist. With evolving professionalization of the sciences during the last decades of the nineteenth century,[45] Lomonosov's

---

[43]   Liubimov, "Lomonosov kak fizik," 3.

[44]   Ibid., 4.

[45]   Helpful micro-studies exploring early efforts at fostering disciplinary cohesion in chemistry are Nathan M. Brooks, "Alexander Butlerov and the Professionalization of Science in Russia," *Russian Review* 57, no. 1 (January

various personae as a chemist, a physicist, or a geologist, to indicate the more obvious appellations, each found its particular historian, but even so, representations of him as an encyclopedic scientist remained the dominant theme in the mythology.

In outlining Lomonosov's overall role in Russian culture, where he again was portrayed as a successor to Peter the Great, Liubimov's disquisition as yet reveals nothing untoward, and quite comfortably reflects the framework of extant Lomonosov biographies. Liubimov's critique of his science cannot, however, have been perceived as anything but utterly disparaging. He prefaced his observations of Lomonosov's qualities as a physicist and chemist with a verdict that reads as a summation of his subject's repute as a scientist: "Lomonosov's name is not connected with any famous discoveries; we are not even able to encounter his name in the history of science."[46] Although this is not precisely delineated in the book, his contention clearly concerned science history in Western Europe,[47] not exclusively in Russia. This stands in contrast to most of the later scholarship, where implicit is the idea that Lomonosov's

---

1998): 10-24; and Michael D. Gordin, "The Heidelberg Circle: German Inflections on the Professionalization of Russian Chemistry in the 1860s," in Michael D. Gordin, Karl Hall, and Alexei Kojevnikov, eds. *Intelligentsia Science: The Russian Century, 1860-1960, Osiris 23* (2008): 23-49.

[46]   Liubimov, "Lomonosov kak fizik," 4. Pekarskii repeats this when opening his attenuated review of Lomonosov's legacy in physics. See his *Istoriia Akademii nauk*, vol. 2, 447-48.

[47]   Ferdinand Hoefer cautioned the readers of his authoritative 1869 study, *Histoire de la Chimie* (2nd ed.), that "among the Russians [of the last century] who acquired fame as chemists was Mikh. Lomonosov, who should not be confused with the poet bearing the same name" (cited in Liubimov, *Zhizn' i trudy Lomonosova*, 60). Established previously is that Lomonosov's scientific writings had a fair amount of exposure during his lifetime in Western Europe. By the nineteenth century, however, they had fallen into obscurity. His literary output, on the other hand, was still occasionally quoted and discussed. For confirmation of this, consult Fomin, *Materialy po bibliografii o Lomonosove*, 123-211. See also N. V. Sokolova's "Kratkii obzor angliiskoi literatury XVIII-XIX vv. o M. V. Lomonosove," in *Lomonosov: sbornik statei*, vol. 7 (Leningrad, 1977), 160-77, which traces Lomonosov in eighteenth- and nineteenth-century English writings.

science, and not simply his cultural meaning, can only be viewed through a Russian prism.

Despite Liubimov's essentialism in his presentation of the process of discovery, his argument is legitimate, at least in how contemporaries understood the merit of Lomonosov's science. It is worth reiterating that prior to Menshutkin's researches, which began in the early years of the twentieth century, only a handful of scientists, Severgin and Liubimov the most prominent among them, even bothered to look with any degree of deliberation at Lomonosov's papers. Since they were seen as archaic, few explications of his dissertations were proffered. While this obviously precluded substantive knowledge of his theories, it in no way reduced the aura of his brilliance as the father of Russian science.

One of the more well-trodden assertions that biographers employed to explain the unfinished nature of much of Lomonosov's scientific work was that the onerous requirements placed on him, by Shuvalov, for example, rendered it impossible for him to devote sufficient time to his scientific duties. That many of his socioprofessional successes were only achievable due to the patronage of Shuvalov is elided. In any case, Liubimov cast a decidedly more critical gaze on this whole notion. He apportioned no space to speculating on what Lomonosov, given the proper honor in his lifetime, might have attempted or contributed; rather, he focused on what he saw as the erudition behind the papers themselves.

Lomonosov may well have expended much of his strength in areas unrelated to the sciences, but Liubimov insisted that these same peripatetic work habits severely undermined his performance within chemistry and physics: "The variety of subjects that he pursued with infinite curiosity caused his attention to move from one field to the other and this did not permit him to remain on a particular investigation of any specific phenomenon; his mind always moved into the area of theory."[48] To be defined as a theorist was, even given the contemporary dismissals in Russia of those engaged in "pure science," not itself an irrevocably sharp criticism.

---

48   Liubimov, "Lomonosov kak fizik," 4.

But if his abilities as a theorist were similarly called into question, might not then his inviolate stature as a scientist be damaged? And Liubimov did indeed dispute the profundity, indeed the value, of Lomonosov's physical and chemical formulations.

Lomonosov's conjectures on the origins of light and color were held up to scrutiny. This was an area that at least since Murav'ev's "Contributions of Lomonosov to Learning" and possibly before it had drawn appreciative responses from Russian observers. Murav'ev was, as will be recalled, quite impressed by the tenacity with which Lomonosov disagreed with Newton's hypotheses. Liubimov was less struck by Lomonosov's doggedness than by his failure to more thoroughly analyze the nature of color.[49] Research on color and light were areas, Liubimov exaggeratedly stressed, which had absorbed a substantial amount of Lomonosov's time and experimental energy. In fact, he had devoted little attention to what can accurately be labeled experimentation, but Liubimov's contention that he had, though futilely, would thus undercut Lomonosov more than it would have otherwise. But even with Lomonosov's strenuous efforts, Liubimov points out, "the inadequate state of chemical knowledge in those days led him onto false paths."[50] Liubimov did admit that "even in our day it is not possible to answer why one body is red, the other yellow, or another color." With the markedly more rudimentary scientific understanding that Lomonosov was

---

[49]   Liubimov's allusions are to Lomonosov's paper *Oration on the Origins of Light, Representing a New Theory of Colors*. Lomonosov's contributions to optics (and to developing a "night-vision telescope") are surveyed thoroughly by Sergei Vavilov in *Mikhail Vasil'evich Lomonosov* (Moscow, 1961), 69-120. Vavilov, a physicist whose specialty was in the area of optics, served as president of the Soviet Academy of Sciences from 1945 until his death in 1951. In addition to writing extensively on various facets of Lomonosov's science, he initiated the publication of the fullest edition of Lomonosov's collected works (the first volume of which was issued in 1950). For an abbreviated discussion of Vavilov's "philosophical and historical writings"—he was, not surprisingly given his interest in light theory, also deeply engaged in the study of Newton—see Alexei Kojevnikov, the author is the same: *Stalin's Great Science: The Times and Adventures of Soviet Physicists* (London: Imperial College Press, 2004), 158-85.

[50]   Liubimov, "Lomonosov kak fizik," 4.

able to bring to his research, "his investigations," it would appear quite inevitably, "did not end in a favorable result."

In a devastating aside, he argued that such colleagues of Lomonosov at the Academy of Sciences as Krafft, Richmann, and Aepinus,[51] while unquestionably less talented than Lomonosov, nonetheless "left their names [or marks] on science" in a lasting manner, unlike him. Liubimov somewhat absolved Lomonosov of responsibility for this outcome, for Krafft, Richmann, and Aepinus, again unlike him, labored in areas in which the ground had long been prepared by those active in European scientific centers. This was not quite accurate: Richmann and Aepinus pursued studies, on electricity especially, which heavily overlapped with Lomonosov's. But to suggest that any of the eighteenth-century academicians, with the exception of Euler, in any way surpassed Lomonosov as a natural philosopher was a heretical judgment.

It was not solely Lomonosov's individual skills that Liubimov deprecated: he more consequentially directed his blame at eighteenth-century Russia's general lack of receptivity to science, quite dismissing the notion that it possessed anything resembling an established scientific tradition. Enfeebling to Lomonosov's potential was the fact that an educated domestic public that might have made up a critical audience for his physical and chemical exercises did not exist. Due to the circumstances in which he spent his most active years, Lomonosov was thus unable to share in the glories of proposing revolutionary hypotheses. On him "lay the task of being an originator; for at that time science was to us like a mysterious temple, and he bore the responsibility of introducing it to his compatriots."[52] Because of this, "He was forced to spend more time teaching than on attempting discoveries." He was severely constrained, then, both by the age in which he lived, although

---

51   Richmann's and Aepinus's activities at the Academy have been noted. Krafft was a natural philosopher and mathematician who left Russian service in 1744. After his return from abroad, Lomonosov worked as a physics adjunct under Krafft's supervision.

52   Liubimov, "Lomonosov kak fizik," 4-5.

this seems occasionally to recede as a causal factor in Liubimov's evaluation, and by geographical circumstances.

Obliged to inculcate a new consciousness among his countrymen, one that would prove more adaptable to developments in the sciences, Lomonosov's duty was no less ambitious than to alter the "mentality of the Russians."[53] Reminding his readers that similar intellectual transformations in Western Europe "took centuries to shape the minds of foreign scientists," Liubimov consistently defended the position that Lomonosov was impelled by Russian conditions to attend almost exclusively to the dissemination of science. This was hardly to be scoffed at, for after all it represented the "highest of missions of a Russian scholar," and one at which Lomonosov excelled. By no means is this contention original to Liubimov; all of Lomonosov's memoirists thought the popularization of science to be one of his singular achievements. But the apparent damage that Liubimov insisted it inflicted on Lomonosov's ability to participate in any theoretical advances was an inference barely alluded to by preceding biographers.

As Liubimov noted, Lomonosov composed many interesting, even fascinating, scientific dissertations, addressing some of the fundamental scientific questions of his era.[54] At first his audience might be lulled into expecting that eulogistic praise is forthcoming, especially as Liubimov places before them one of Euler's [presumed] critiques of Lomonosov's corpuscular papers.[55] Having examined Lomonosov's suppositions, Euler expressed his pleasure at reading them, for what he has seen so far demonstrates that Lomonosov "is in possession of a fortunate capacity for delineating phenomena in physics and chemistry." This nicely, though also very vaguely, buttresses the notion of Lomonosov's prowess. Euler's evaluation was an established resource. Even so, it is evident that Liubimov

---

53   Ibid., 5.

54   Ibid.

55   The treatises assessed by Euler were *A Dissertation on the Action of Chemical Solvents in General* and *[Physical] Meditations on the Cause of Heat and Cold.*

resorted to it in order to allow his less palatable conclusions to get an airing.

There is little doubt that Liubimov granted no credence to the idea that Lomonosov was a Russian Newton, Boyle, or Franklin. For although "few of his contemporaries understood the happenings of nature more deeply or more lucidly than he," it also must be acknowledged that "Lomonosov was never able to introduce any startling new facts into science." This is explained by the primary deficiency in Lomonosov's scientific skills, a weakness that from the first severely limited his potential: "Lomonosov was not a mathematician; his theories carried a purely physical character." This topic has been broached more than once, but it bears returning to, for however much Lomonosov may have believed that mathematics was necessary to chemistry and physics, he did not partake in the eighteenth-century "revolution" in mathematical analysis.

In his *Elements of Mathematical Chemistry* (*Elementy matematicheskoi khimii*, 1741), Lomonosov paid obeisance to the wonders that might be wrought by fusing mathematics and the sciences:

> That light which mathematics is able throw on chemistry [*spagiricheskaia nauka*], may be foreseen by those who know its secrets and also know the main natural sciences, which are perfected by the use of mathematics, such as hydraulics, aerometry, optics, and so forth. Everything that used to be dark, dubious, and unsure in these sciences, mathematics has made distinct, reliable and obvious.[56]

But despite such soaring rhetoric, Lomonosov's essays were largely devoid of any such intermingling of method and practice. Employing mathematical terminology and giving his dissertations mathematical-sounding titles very nearly approximates the extent of his exploration of the subject.[57] Liubimov clearly perceived

---

[56]   Lomonosov, *PSS*, vol. 1, 75.

[57]   Valentin Boss contends that this is one of the more evident end products of his studies in Germany: "It is one of the traits he borrowed from Wolff;

how crippling this was to Lomonosov' work, and, with restraint, pointed it out.

Describing the state of the physical sciences in eighteenth-century Europe, Liubimov rigidly divided disparate European scientific circles into two polarized halves: the adherents of Descartes and those of Newton. But in this contest, the powerful logic of Newtonian philosophy decisively routed any alternative approaches toward understanding nature. Liubimov was a bit too stark in demarcating Newton's and Descartes's followers, but he did accurately reflect the apparent triumph of Newtonian ideas in European intellectual life. "At the end of the [eighteenth] century Newton's victory was complete," Alexandre Koyré remarked in attempting to fathom the dimensions of the "scientific revolution," and "the Newtonian God reigned supreme in the infinite void of absolute space in which the force of universal attraction linked together the atomically structured bodies of the immense universe and made them move around in accordance with strict mathematical laws."[58]

---

his expositions are clearly and arithmetically ordered, but it was a purely formal characteristic that had nothing in common with mathematical analysis in the Newtonian sense." Boss, *Newton and Russia*, 180. See also Leicester, *Lomonosov on the Corpuscular Theory*, 12-13.

[58] Alexandre Koyré, *From the Closed World to the Infinite Universe* (Baltimore: Johns Hopkins University Press, 1957), 274. Koyré qualified this rather absolute profession by still allowing a somewhat heterogeneous character for the Newtonian synthesis. Indeed, his conclusions are not too distant from current historiography, as it was, for example, recently enunciated by Peter Dear: "it is worth observing that the story was not to be a simple one of Newtonian 'truth' beating out Cartesian 'romance' (as some critics liked to characterize Descartes's mechanical universe). The complexity and interweaving of arguments, mathematical, metaphysical and experimental, meant that … what counted as 'Newtonianism' was in many ways quite different from what Newton himself had believed and argued. The 'Newtonianism' of the later eighteenth century was itself a hybrid of Newton's, Descartes's, Leibniz's and many other people's work and ideas." Peter Dear, *Revolutionizing the Sciences: European Knowledge and Its Ambitions* (Princeton: Princeton University Press, 2001), 167. Although in Liubimov's essay a more purist vision of Newtonianism is conveyed, he also admits a degree of heterodoxy.

Even given what appears to have been the inevitable ascendancy of this new system of knowledge, Liubimov carefully explained that it must be remembered that in the eighteenth century only "slowly did the ideas of Newton penetrate into science," for the rather fundamental reason that to appreciate, let alone comprehend, Newton's ideas required an "extensive understanding of mathematics."[59] Therefore, "in physical investigations the majority of scientists," who did not yet grasp the significance of mathematics, "followed the path fixed by Descartes."

Liubimov explains that not surprisingly, in light of Lomonosov's only rudimentary grasp of the advanced mathematics of his time, the "works of Newton did not have a great influence" on his thinking, and perhaps more provocatively states that "in his views on nature he was purely Cartesian."[60] Whether or not Lomonosov's opinions were strictly along these lines, and Liubimov reiterated his point lest his readers not fully appreciate its import, is debatable. Lomonosov was convinced by Cartesian mechanics,[61] at any rate far more than he was by Newton, whose theories he largely disdained. But the more interesting point is that with the eventual triumph of Newtonianism across Europe, Liubimov profoundly called into question Lomonosov's scientific worldview.[62] Framed

---

[59]   Liubimov, "Lomonosov kak fizik," 6.

[60]   For as Liubimov again assured his readers, "Lomonosov was not a mathematician" ("Lomonosov ne byl matematikom"). Ibid., 7.

[61]   Reflecting historically on Descartes's impact on natural philosophy, Lomonosov, in his preface to the *Volfianskaia eksperimental'naia fizika* (Lomonosov, *PSS*, vol. 1, 423), paid ornate tribute to Cartesian natural philosophy for having "dared" to challenge the Aristotelian dominance, and to have "disproved" its ideas.

[62]   Liubimov examined ("Lomonosov kak fizik," 13-31) several of Lomonosov's dissertations, both those published initially in Latin and the more "accessible" Russian language ones, including *Meditations on the Cause of Heat and Cold; Oration on the Origins of Light, Representing a New Theory of Colors; Discourse about Air Phenomena, Caused by Electricity; Discourse on Greater Accuracy of the Sea Route;* and *The Appearance of Venus Before the Sun.* He also introduces the relatively more obscure *An Attempt at a Theory of the Elastic Force of Air (Opyt teorii uprugoi sily vozdukha*—issued in *Novi*

in this seemingly binary fashion, it would appear that Lomonosov was on wrong side of an epic divide.

That Cartesianism had an intellectually injurious effect on Lomonosov is most evident in his refusal to accept, or even properly appreciate, the notion of attraction at a distance.[63] This resulted from the fact that, approximating Descartes, he repudiated the idea of a vacuum in space. Lomonosov was thoroughly wedded to a mechanical/corpuscular view on cosmological questions, and that space could exist without matter was utterly contradictory to his reasoning. Lomonosov's refusal to embrace gravitational theory was one shared by some illustrious scientific figures. After all, even Euler, as Liubimov commented, referred to Newton's essential conception as, "obscura attractio quorundam Anglorum." However much Liubimov may have allowed that this was hardly an unusual position for the times, it was, nonetheless, a devastating verdict on Lomonosov.

But while Cartesianism might, in Liubimov's view, have been utterly bested by Newtonian prescriptions, Descartes still personified an acceptable stage in the linear outline of scientific progress that he sketched out. Wolffian natural philosophy was, on the other hand, hopelessly misguided and scarcely deserving of serious consideration. That Wolff himself was deeply influenced by Cartesian mechanics, perhaps as much as he was by Leibnizianism,

---

Commentarii in 1750). For this last treatise, which was first translated in its entirety into Russian by Boris Menshutkin, see Lomonosov, *PSS*, vol. 2, 105-39, 653-57. Liubimov did concede that Lomonosov's navigational paper in particular conveyed "practical information," but as for the rest, he by and large dismissed them as "without doubt of only historical significance." Liubimov, "Lomonosov kak fizik," 16. While the latter remark concerns Lomonosov's essay on heat and cold, analogous assessments characterize each of the essays. Throughout his analysis Liubimov is apparently most intent on delineating the Cartesian and quasi-Cartesian makeup of Lomonosov's theorizing. However, as is stressed by Liubimov when discussing Lomonosov's optical investigations, in which his dependence on Descartes's mechanical hypotheses on the nature of color and light appear pronounced, Lomonosov did add his own, often original, not to say prescient, thoughts.

[63]   Ibid., 12-13.

does not enter into Liubimov's analysis. Because of Lomonosov's association with Wolff, he had, nevertheless, to at least ponder the irksome question of Wolff's sway over him.

The mathematical issue reappeared as a related and pressing concern. While far from enthusiastic about Lomonosov's analytical proficiencies, Liubimov exculpated him, without presenting any substantive arguments, from direct contamination by Wolff. As he put it, although "Lomonosov studied in Germany and attended courses with the famous mathematician and physicist Wolff ... German scientists were less of an influence on him than were the French."[64] The reason was quite uncomplicated: "The clear mind of Lomonosov did not submit itself to the formulaic models that characterized the writings of German scientists, especially Wolff."[65]

Eventually Lomonosov did largely abandon Wolff's formal method of employing mechanical structures for demonstrating proofs in his disquisitions, so Liubimov's statement is in part correct, at least for papers that Lomonosov composed after he had matured and found his own style. This does not treat the more important element, however, of whether Lomonosov discarded Wolff's basic methodological assumptions. A perusal of Lomonosov's corpuscular treatises indicates quite the opposite. In any case, Liubimov's investigation of the link between Lomonosov and Wolff is cursory, and offers little more than a curt rejection of any salient intellectual association between them following Lomonosov's return to St. Petersburg. Despite Descartes having been overtaken by Newton, he still represented, it seems, a more enlightened, and a more scientifically sophisticated, approach than Wolff's.[66]

---

64   Ibid., 31.

65   Ibid., 32-33. He conceded Wolff's "direct influence" over only one paper: *O vol'nom dvizhenii vozdukha,...*" (*About the Free Movement of Air...*," first published in *Novi Commentarii* in 1750). See Lomonosov, *PSS*, vol. 1, 315-33, 564-66. It is in this treatise, Liubimov accurately notes, that Wolff's style of exposition, with arguments by use of "corollaries," "theorems," and "definitions," is plainly evident.

66   As reported in "Liubimov," *Entsiklopedicheskii slovar'*, 209, later in life (1886) Liubimov translated some of Descartes's writings. This probably indicates a continuing respect for Descartes's place in the history of science.

In the area of equivalence with Franklin, Liubimov granted little credence on the more excessive claims made on Lomonosov's behalf. Although he appears to have been reticent to utterly overturn the notion of Lomonosov anticipating Franklin, he did aver that "many have stood up to convey hypotheses of lighting and electricity flashes…. But Franklin was the first … to prove from the flash he extracted from the skies that [this spark] has all of the properties of the spark of electricity."[67] He gave Lomonosov credit for his own speculations, which roughly paralleled those of Franklin, and recognized that in some of his conjectures he may have even exceeded Franklin's reach, but he was quite explicit in stating that Lomonosov was inspired to do his own experiments after finding out about Franklin's.[68] That having been said, Lomonosov was not aware of all the details of Franklin's work, so his theorizing still had an aura of originality. As was every other biographer of Lomonosov, in his inquiry into his electrical work Liubimov was much taken with the events surrounding the death of Richmann. Like Pogodin, he reprinted in full Lomonosov's letter to Shuvalov describing the incident, along with his hope for future scientific progress in Russia.[69]

Liubimov's appraisal of Lomonosov need not be interpreted as either unremittingly, or even primarily, condemnatory. While Liubimov did not support the notion that Lomonosov possessed a prodigious scientific mind, he granted that his contributions have "for us a more important meaning as one of the brightest pages in the history of Russian education."[70] Such a determination, at least superficially, does not differ from Pogodin's, and Liubimov broadly imitated his fellow memoirists by writing, "his love for

---

[67]   Liubimov, "Lomonosov kak fizik," 20-21.

[68]   Lomonosov's own references to Franklin's work were elucidated earlier. Liubimov may also, in remarking specifically about Franklin's kite experiment, have had in mind a notice in *Sanktpeterburgskie vedomosti*, no. 47 (1752), which, very generally, brought to "public attention" Franklin's research.

[69]   Liubimov, "Lomonosov kak fizik," 25-27.

[70]   Ibid., 6.

science, and his wish to disseminate it in our fatherland—that is the predominant essence of Lomonosov."[71] Furthermore, Moscow University itself, which has become so "closely connected with the name of Lomonosov, that it is, it seems, possible to state that his undertakings have not vanished,"[72] as well as his initiative in naming the first Russian professors to it,[73] must, Liubimov declaimed to his audience, be counted among his singular bequests to his country.

To repeat a consequential point, virtually no attention has been paid by Russian and Soviet historians of science to Liubimov's work. But might not the dedication of so much space to what is in essence a single example be construed as an effort in overt deconstruction, or more precisely an attempt to demolish an icon? Perhaps, though this was not the rationale for its use. Rather, I would contend that aberrations within or from myth can reveal

---

[71] Ibid., 34. Again echoing his fellow biographers, he repeats Lomonosov's "deathbed" lamentation to Staehlin.

[72] Ibid., 35.

[73] Lomonosov's apparent patronage of the first Russian professors at Moscow University became part of the imagery around his founding of the university. Much of this has to do with the national makeup of its first academic staff, which included far more Russians than that of the Academy of Sciences, thus leading to the supposition that Lomonosov was intent on furthering the careers of his countrymen. Nikolai Popovskii, who was part of the first cadre of professors attached to the university, is described within the historiography as his protégé. Popovskii, who had aroused immense controversy with his translation of Pope's *Essay on Man*, was Lomonosov's student at the Academy of Sciences, and also enjoyed the support of Ivan Shuvalov. A professor of eloquence at Moscow University, he delivered one of the inaugural addresses at the university's opening. Unfortunately, Popovskii died relatively early in his professional life, in 1760, predeceasing Lomonosov by five years. Between this and the premature death of his chemistry student Klement'ev, Lomonosov's chances of finding a successor at the Academy of Sciences were presumably dashed. It would seem that similar circumstances left him without an heir, at least one of real potential, at Moscow University. L. B. Modzalevskii, "Lomonosov i ego uchenik Popovskii (o literaturnoi preemstvennosti)," *XVIII vek* 3 (1958): 111-69, makes an effective argument for Popovskii's reliance on Lomonosov's literary tutelage and bureaucratic aid. For a brief record of Popovskii's association with Moscow University, see Shevyrev, *Istoriia Imp. Moskovskogo universiteta*, 26-30.

as much as the received wisdom can.[74] Liubimov's essay speaks far
more about what constituted the prevailing imagery of Lomonosov
in mid-nineteenth century Russia than do the myriad writings
that simply echo the existing mythology. There is, not to put too
strong a gloss on it, a palpable sense of indignation in his appraisal
at perceived distortions in the historical record of Lomonosov's
science, along with tremendous respect for a natural philosopher
who attempted much and who emerged as a commanding symbol
for later generations.

Omitted from this chapter has been a detailed review of what
has received more enthusiastic coverage in the literature than any
other single episode or biographical item prior to Menshutkin's
arrival on the scene: the 1865 Lomonosov Jubilee. It is widely held, in
my view misleadingly, that it was during 1865 and in the immediately
succeeding period that research on Lomonosov, both his scientific
and literary halves, was first raised to a higher, more complex plane.[75]
That year marked the hundredth anniversary of Lomonosov's
death, and to honor his role in Russian culture, ceremonies were

---

[74]   Or to quote Irina Reyfman, "At least as helpful in reconstructing an epoch's
         collective self-image are conscious deviations from common views."
         Reyfman, *Trediakovsky*, 1. On the usefulness of examining "systematic
         omissions" in the portrayal of iconic scientists, see Pnina G. Abir-Am,
         "How Scientists View Their Heroes: Some Remarks on the Mechanism of
         Myth Construction," *Journal of the History of Biology* 15, no. 2 (Summer 1982):
         281-315; Abir-Am and Elliot, *Commemorative Practices in Science*; and
         Shortland and Yeo, *Telling Lives in Science*.

[75]   To varying degrees exemplifying this thesis as well as offering introductions
         to the jubilee and its related literature are P. N. Berkov, "Lomonosovskii
         iubilei 1865 g.," in *Lomonosov: sbornik statei*, vol. 2, 216-47; B. F. Egorov,
         "Lomonosovskii iubilei 1865 g.," in *M. V. Lomonosov i russkaia kul'tura:
         tezisy dokladov konferentsii, posviashchennoi 275-letiiu so dnia rozhdeniia
         M. V. Lomonosova (28-29 noiabriia 1986 g.)* (Tartu, 1986), 56-59; V. P. Lystsov,
         M. V. Lomonosov v russkoi istoriografii 1860-1870-x godov (Voronezh, 1992);
         Radovskii, *Lomonosov i Akademiia nauk*, 231-42; Solov'ev and Ushakova,
         Otrazhenie estestvennonauchnykh trudov Lomonosova v russkoi literature,
         57-84. At least as concerns investigations into Lomonosov's scientific legacy,
         Alexander Vucinich regards the jubilee period with a more jaundiced eye.
         See his *Science in Russian Culture: A History to 1860*, 401-02; and idem, *Science
         in Russian Culture: 1861-1917*, 69-70.

organized in more than twenty locations throughout the Russian Empire, with the main celebrations taking place at the Academy of Sciences in St. Petersburg.[76] The most visible historical trace of

---

76   Based on a search of the contemporary press, Berkov tentatively proposed that in addition to St. Petersburg and Moscow, twenty-three other cites, towns, or villages held Lomonosov celebrations. See his "Lomonosovskii iubilei," 235. For a remarkably detailed, and reverential, description of the 1865 Lomonosov Jubilee ceremonies in St. Petersburg, which lasted three days, see P. I. Mel'nikov, *Opisanie prazdnestva, byvshago v S.-Peterburge 6-9 aprelia 1865 g. po sluchaiu stoletniago iubileia Lomonosova* (St. Petersburg, 1865). The events consisted of church services, dinners punctuated by lavish toasts, musical and dramatic interludes, speeches by leading church and state officials, lectures by members of the Academy of Sciences, the unveilings of paintings and busts of Lomonosov, and so forth. Various descendants of Lomonosov, along with assorted Shuvalovs and Vorontsovs, heirs of his main patrons, were in attendance throughout. Present also were some of the leading writers and critics of the time, among them: Fedor Dostoevskii, Ivan Goncharov, Apollon Maikov, Pavel Annenkov, and Fedor Tiutchev. A stellar array of Lomonosov's biographers or chroniclers, including Ia. K Grot, V. I. Lamanskii, Sukhomlinov, and Perevoshchikov, were involved in the proceedings. Finally, Dmitrii Mendeleev, soon to be Russia's most famous chemist, is likewise on record as having come to the celebrations (see ibid., 39-46, for a partial listing of those who attended the jubilee). Marcus Levitt's dissection of the 1880 Pushkin festivities, *Russian Literary Politics*, throws deserved light on the importance of jubilee culture in Russian intellectual, social, and political life. His study offers more than its subject implies, for he also takes the reader into Soviet-era Pushkin imagery. When remarking of the 1880 celebrations, however, that "Never before had so many of Russian's leading novelists, poets, playwrights, editors and publishers, critics and reporters, educators and scholars, actors, artists and musicians, city and state officials—so many of the nation's cultural leaders and opinion makers—gathered together in one place to salute Russian literature" (ibid., 1), Levitt rather overstates his case. He too peremptorily dismisses the Lomonosov Jubilee, which richly solemnized both Lomonosov's literary and scientific activities, as an "in-house event" put on by the Academy of Sciences and Moscow University (ibid., 35). The political, scientific, and nationalist goals which not only encompassed Polish and German objectives, but in time even American ones, that were, and are, "embedded" in the commemoration of Copernicus, an archetypal hero of science, are explored by Owen Gingerich in "The Copernican Quinquecentennial and its Predecessors: Historical Insights and National Agendas," in Abir-Am and Elliot, *Commemorative Practices in Science*, 37-60. Gingerich's expansive approach towards the organization of scientific remembrance can be utilized in analyzing the institutional fashioning

the jubilee is found in the unprecedented outpouring of studies, numbering in the hundreds, which were published in conjunction with the occasion.[77] Although any selective enumeration of these works is inevitably highly subjective, it might be asserted that the fundamental publications to emerge from this deluge were the massive documentary compilations from P. S. Biliarskii and A. Kunik.[78]

Other much-utilized studies released in 1865 include smaller collections of primary sources from Pekarskii and V. I. Lamanskii. Ia. K. Grot drafted a more monographic volume, dealing almost exclusively with Lomonosov's association, narrowly defined, with the Academy of Sciences.[79] Kunik, Lamanskii, and Pekarskii were primarily historians, Biliarskii and Grot philologists, and each of them already was or would in time become a member of the Academy of Sciences. Through their jubilee writings these scholars were determinedly aiming to enshrine in the "public arena"

---

of a scientist's historical persona generally. John L. Heilbron advances a compelling case for the success of centennial commemorations, nourished by the growth of professionalization in the sciences, to decisively enshrine scientists alongside "the great men, *i grandi*, the heroes of history" in the closing decades of the nineteenth century ("Galvani, Volta, and the Uses of Centennials," in *Luigi Galvani International Workshop: Proceedings, Bologna, 9 October 1998*, ed. Marco Bresadola and Giuliano Pancaldi [Bologna: University of Bologna 1999], 17-32). Although Russia is not included in its case studies, John R. Gillis, ed., *Commemorations: The Politics of National Identity* (Princeton: Princeton University Press, 1994), affords an interesting comparative survey of the construction of memory and national identity.

[77] See the following guides to this literature: Fomin, *Materialy po bibliografii o Lomonosove*; V. I. Mezhov, *Iubelei Lomonosova, Karamzina i Krylova: bibliograficheskii ukazatel' knig i statei, vyshedshikh po povodu iubileev* (St. Petersburg, 1871); and S. I. Ponomarev, ed., *Materialy dlia bibliografii o Lomonosove* (St. Petersburg, 1872).

[78] Biliarskii, *Materialy dlia biografii Lomonosova*; Kunik, *Sbornik materialov*, 2 parts.

[79] Pekarskii, *Dopolnitel'nye izvestiia*; V. I. Lamanskii, *Lomonosov i Peterburgskaia Akademiia nauk. Materialy k stoletnei pamiati ego 1765-1865 goda, aprelia 4-go dnia* (St. Petersburg, 1865); Ia. K. Grot, *Ocherk akademicheskoi deiatel'nosti Lomonosova* (St. Petersburg, 1865). Grot's composition (a comparatively slender fifty-eight pages), is a rather conventional, though still very serviceable, memoir of Lomonosov's activities at the Academy.

recognition of the Academy's inestimable contributions to Russian progress. As we have seen, a common method of attempting this was to blend the Academy of Science's history with that of its most glorious emblem, Lomonosov.

My brief comments hardly do justice to either the richness of their content or to the productive uses to which these works are still put. Previous chapters in this volume reveal my own reliance on Pekarskii,[80] both his 1865 and 1873 studies, and to a lesser extent on Biliarskii and Kunik. What distinguished these publications was not, however, their novel readings of Lomonosov's science, but rather the easy accessibility they provided to a vast array of hitherto unpublished or dispersed materials related to his professional activities. What they did not do is to fundamentally reconfigure the representations of Lomonosov as the father of Russian science. Additional mention need be made of two collections of articles that were generated from the jubilee gatherings held at Moscow and Khar'kov Universities.[81] Largely, and expectedly, laudatory in nature, these volumes contain addresses by some of the more prominent historians, scientists, and literary specialists attached to those schools.

Menshutkin, whose pronouncements about Lomonosov in most instances have achieved the status of sacred writ, by and large refrained from attaching any crucial significance to the 1865 jubilee. While this may have partially been the result of his drive to establish himself as *the* pioneer in the study of Lomonosov, much of his palpable indifference to the interpretive value of the so-called jubilee literature undoubtedly stemmed from the fact that

---

[80] Biographical information on Pekarskii can be found in M. V. Mashkova, *P. P. Pekarskii (1827-1872): kratkii ocherk zhizni i deiatel'nosti* (Moscow, 1957). For a survey of Pekarskii's depiction(s) of Lomonosov, see V. P. Lystsov, *Zhizn' i deiatel'nost' M. V. Lomonosova v osveshchenii P. P. Pekarskogo* (Voronezh, 1993). As with Lystsov's earlier publications on Lomonosov, however, his ideological sentiments restrict the value of his labors.

[81] *Prazdnovanie stoletnei godovshchiny Lomonosova 4-go aprelia 1765-1865 g. Imperatorskim Moskovskim universitetom;* and *Pamiati Lomonosova. 6-go aprelia 1865 goda* (Khar'kov, 1865).

he did not espy any new or conceptually ambitious evaluations of
Lomonosov's science within them.[82] In the matter of the writings
emerging from the jubilee period, however, Menshutkin's implicit
disavowals of their scientific consequence had only an indirect
impact on an expansive later historiography.

That there was a noticeable diminution in the accrual of new
elements and vigorous positive reevaluations brought to Lomonosov's scientific biography in the decades since its encounter with
Pushkin is unequivocal. This meant, of course, that Lomonosov's
weighty symbolic presence in Russian culture was in danger of lan-
guishing. While Pogodin's memoir was a spirited advocacy of Lomonosov's legacy, its main innovation was in its attempt to link his
renown more surely to the contemporary fate of Moscow University and the Academy of Sciences. While this supplemented Lomonosov's broader status in the historical discourse, as well as providing
needed prestige to the aforementioned institutions, it contributed
little to the portrayals of his more purely scientific exploits.

As for Liubimov, his intriguing essay plainly did not nourish
the mythology; rather it could, especially if followed by like
receptions, have signified the beginnings of its inexorable decline.
Lomonosov's position in Russian thought was still secure enough
to fend off Liubimov's quite solitary challenge; whether it could
withstand multiple such threats in the future seemed problematic.
Then in the first decades of the last century the myth of Lomonosov,
and the related imagery of Russian science, were inestimably
invigorated by the work of Boris Menshutkin. He would, for the first
time since the casting of Lomonosov's biography in the eighteenth
century, not only refine or modify the idea of Lomonosov as the first
and most splendid of Russian scientists, but substantially expand
its reach.

---

[82]   As will be shown in the succeeding chapter, a limited exception may have
         been extended by Menshutkin to Anton Budilovich's *M. V. Lomonosov kak
         naturalist i filolog* (St. Petersburg, 1869).

Chapter 5

Boris Menshutkin
and the "Rediscovery" of Lomonosov

8 November 1911 marked the two hundredth anniversary of Lomonosov's birth,[1] and the occasion witnessed another round of jubilee ceremonies, with the principal assembly convening that evening at the Academy of Sciences in St. Petersburg.[2] Although commemorative occasions would continue to be held at periodic intervals in ensuing decades to note both Lomonosov's

---

[1] Although Lomonosov's exact date of birth is unknown, based on the suppositions of M. I. Sukhomlinov (see his "K biografii Lomonosova," in *Izvestiia Otdeleniia russkogo iazyka i slovesnosti Imperatorskoi Akademii nauk* 1, book 4 [St. Petersburg, 1896], 782-83) 8 November 1711 has been widely accepted as "official." A. I. Andreev traces the background of investigations into Lomonosov's date of birth in "O date rozhdeniia Lomonosova," in *Lomonosov: sbornik statei*, vol. 3, 364-369. Having largely undermined Sukhomlinov's case, which was built on decidedly inconclusive evidence, he leaves the question unresolved.

[2] The activities of the Academy of Sciences in preparing for the Lomonosov Jubilee of 1911 began in earnest in 1909 with the creation of a commission to plan the festivities. E. S. Kuliabko, "Lomonosovskii iubilei 1911g.," in Berkov, *Literaturnoe tvorchestvo Lomonosova*, 300-12 and Radovskii, *Lomonosov i Akademiia nauk*, 249-59, offer thorough reviews of the jubilee's design and execution. For a contemporaneous account, see also *Lomonosovskiia torzhestva. (Bibliograficheskaia zametka)*, 88-105, in *Pamiati M. V. Lomonosova. Sbornik statei k dvukhsotletiiu so dnia rozhdeniia Lomonosova* (St. Petersburg, 1911), 88-105.

birth and death (for example in 1915, 1936, 1961, 1965, and 1986),[3] the 1911 jubilee was particularly significant due to its enormous success in furthering Lomonosov's reputation as a scientist. During the 1865 celebrations, Lomonosov's role had largely been subsumed to that of the Academy of Sciences; or rather his accomplishments were depicted as inseparable from those of the Academy. At any rate, his scientific biography at that time had received no accretions of new views that in any way altered the then-prevalent image, albeit a broadly drawn one, of his scientific exploits on behalf of his country.

As for Soviet-era commemorations, these were bereft of any sense of a living myth of Lomonosov, as his biography was employed exclusively to buttress national pride. This had, of course, also been crucial to nearly all the pre-revolutionary accounts of his life. In more recent decades, however, this element taken on an utterly

---

[3]   These dates, of course, coincide with the dates of Lomonosov's birth and death, 1711 and 1765 respectively. Nikolai Krementsov maintains that commemorations, at least during the Soviet period, were usually held at twenty-five year intervals, with the addendum that "unusual figures generally signal unusual occasions." Krementsov, *Stalinist Science* (Princeton: Princeton University Press, 1997), 326. As is evident from the above dates, "unusual occasions" seem to have characterized the evolution of Lomonosov jubilee culture. The past few decades have seen a proliferation of jubilee celebrations in many countries (for more on this apparent "commemorative mania," see Pnina G. Abir-Am, introduction to Abir-Am and Elliot, *Commemorative Practices in Science*, 1-33), but perhaps nowhere were they more prevalent than in the Soviet Union. The distinction between solemnizing a revered figure's birth or death—such as the decision undertaken in 1955 (in the run-up to de-Stalinization) to focus the main celebrations of Lenin on the date of his birth, rather than as before on the day of his death (on this consult Tumarkin, *Lenin Lives! The Lenin Cult in Soviet Union* [Enlarged edition, Cambridge, MA: Harvard University Press, 1997], 257-58) - seems not to have affected the scheduling of festivities extolling either Lomonosov or, for example, Pushkin, another figure subjected to intense idolization during the Soviet era. For an exploration of the Pushkin myth at its most extreme—and hagiographic—expanse, during the 1937 jubilee commemorating his death, see Iurii Molok, *Pushkin v 1937 godu: materialy i issledovaniia po ikonografii* (Moscow, 2000).

proscribed quality.[4] By contrast the 1911 jubilee, although directed

---

[4]   As perceptively noted by Krementsov when discussing the ideological canonization, beginning in the 1930s, of scientists, specifically those who could be attached to particular disciplines, "Celebrations of an event in a founding father's life, such as birth, death, or publication of an important work, were used to stage public demonstrations—sanctioned, of course, by party authorities and signifying party approval of not only the founding father, but also the discipline or institution commemorating the jubilee. The very list of recognized founding fathers and their essential characteristics emphasized in numerous glorifications, then, reflected the image of science and the scientists endorsed by the party authorities." Krementsov, *Stalinist Science*, 222. This complete annexation of commemorative culture in the sciences to the party-state, as distinguished from events that previously had been largely under the purview of a particular institution or discipline, while they were hardly free from often heavy-handed or clumsy regime interference—was accomplished rather easily, for it was "simplified and facilitated by the cult of the 'founders of the party,' Marx, Engels, Lenin, and Stalin, that permeated the Bolshevik political culture. Soviet scientists included these sacral ideological authorities in their own pantheon of Great Scientists, spreading the authority of party founders over their own 'founding fathers'. The Party apparatus, in turn, recognized the authority of Great Scientists, establishing special prizes for scientific research named after founding fathers, celebrating their various anniversaries, and giving names to scientific institutions." The effect of such crude politicization on jubilee culture was to eventually render it meaningless. This process quickly extended to internal processes within the disciplines themselves, and was employed by scientists to protect and extend their own domains, for as Krementsov emphasizes, "Any criticism of the founding father's research was regarded as an assault on an exalted ideological authority. Their legacies were invoked to legitimate almost every new approach within these disciplines; many scientists claimed that their work directly originated from a founding father's research. Their authority was also used to contrast 'native' and foreign science in the patriotic campaigns or to validate the 'practicality of science'" (ibid., 50-51). Krementsov's discerning study is somewhat marred, however, by his conviction that scientific jubilees as a force in the nation's scientific life emerged mainly in the 1930s (ibid., 52), which of course reflects a tendency when approaching the history of Russian and Soviet science to separate rather too mechanically, and too sharply, between what was Soviet and what was Russian, without admitting the continuities. For more on the study of the history of science in the Soviet Union, with an emphasis on the more desultory effects on the discipline of the need to satisfy the shifting demands of Stalinist culture, see Loren R. Graham, "The Birth, Withering, and Rebirth of Russian History of Science," *Kritika* 2, no. 2 (Spring 2001): 329-40; idem, *The Soviet Academy of Sciences and the Communist Party, 1927- 1932* (Princeton:

by the Academy of Sciences, was one in which the Russian scientific
community as a whole exhibited an extraordinary amount of vigor
in examining its own past as well as in arguing for its contemporary relevance, while at the same time fulsomely honoring
Lomonosov.

After the profusion of studies on Lomonosov during the 1860s
and 1870s, there had been a noticeable diminution of new works
in subsequent years.[5] It would seem that with literary devotions

---

Princeton University Press, 1967); David Joravsky, *Soviet Marxism and
Natural Science, 1917-1932* (New York: Columbia University Press, 1961),
215-314; idem, "Soviet Views on the History of Science," *ISIS* 46, no. 143
(1955): 3-13; L. V. Levshin, *Sergei Ivanovich Vavilov, 1891-1951* (Moscow,
2003), 160-358; James T. Andrews, *Science for the Masses: The Bolshevik State,
Public Science, and the Popular Imagination in Soviet Russia, 1917-1934* (College
Station, TX: Texas A & M University Press, 2003), 154-76; Vera Tolz, *Russian
Academicians and the Revolution: Combining Professionalism and Politics* (New
York: St Martin's Press, 1997); and Alexander Vucinich, "Soviet Marxism
and the History of Science," *Russian Review* 41, no. 2 (April 1982): 123-43.
Alexei B. Kojevnikov's chapters on Sergei Vavilov ("President of Stalin's
Academy"), and Petr Kapitsa ("Piotr Kapitza and Stalin's Government:
A Study in Moral Choice"), in *Stalin's Great Science,* are also instructive.
For earlier interactions between a frequently supportive new Soviet state
and scholars interested in the history of science and technology, notable
is V. M. Orel and G. I. Smagina, eds., *Komissiia po istorii znanii 1921-1932
gg. Iz istorii organizatsii istoriko-nauchnykh issledovanii v Akademii nauk:
sbornik dokumentov* (St. Petersburg, 2003). This compendium points to the
importance early Soviet scientists, historians, and political figures attached
to re-examining and publicizing Lomonosov's legacy. Researchers affiliated
with the Institute of the History of Science and Technology of the Russian
Academy of Sciences have produced a range of works over the past two
decades on the interaction between Soviet-style communism and the
epistemological roots of the history of science. Many of their writings also
deal with the fate of individual scientists and disciplines. For a sampling
of relevant studies, peruse the more recent issues of the Institute's journal,
Voprosy istoriii estestvoznaniia i tekhniki, which has pertinent pieces in nearly
every number.

[5]   This is relative comparison, for large numbers of items continued to be issued
dealing with Lomonosov throughout the last decades of the nineteenth
century (Fomin, *Materialy po bibliografii o Lomonosove*). This includes quite
well-researched full-scale biographies, such as V. I. Lamanskii, *Mikhail
Vasil'evich Lomonosov: biograficheskii ocherk* (reprint, St. Petersburg, 1883); and
especially A. I. L'vovitch-Kostritsa, *M. V. Lomonosov: ego zhizn', nauchnaia,*

such as that displayed during the 1865 jubilee more and more restricted to such commemorative occasions,[6] those interested in Lomonosov would have to await the arrival of another anniversary. While the 1911 jubilee marked a crucial juncture in the evolution of Lomonosov's reputation as a scientist, the path toward it was thematically somewhat prepared by the issuance in 1901 of a collection of articles on the history of chemistry in Russia.[7] More specifically, this work, inspired by the efforts of the chemist and Moscow University professor V. V. Markovnikov, was designed to call attention to the one hundred and fiftieth anniversary (1898) of Russia's first chemical laboratory. It surveyed efforts across the Russian Empire over the previous century and a half to develop chemical laboratories.

Lomonosov's role as both the founder of the first chemical laboratory in Russia and an inspiration for later generations of chemists and scientists in related fields was awarded wide coverage at the meetings held in Moscow (from 2-4 January 1900) under the auspices of the Chemistry Section of the Society of Admirers of Natural Science, Anthropology, and Geography (which was headed by Markovkinov), out of which the aforementioned volume emerged.[8] Markovnikov's appeal to Russian scientists and to the

---

literaturnaia i obshchestvennaia deiatel'nost' (St. Petersburg, 1892). L'vovich-Kostritsa's entry incorporates a fair amount of the documentary evidence on Lomonosov's work at the Academy that was published during the 1860s and 1870s.

[6] This phenomenon can also be seen in publication statistics concerning Pushkin, for which see Levitt, *Russian Literary Politics*. On the vast expansion of the Russian reading public in late Imperial Russia, which, although the connection is not explicitly discussed by the author, can only have contributed greatly to the strength of jubilee culture, see Jeffrey Brooks, *When Russia Learned to Read: Literacy and Popular Literature, 1861-1917* (Princeton: Princeton University Press, 1985), 295-352.

[7] *Lomonosovskii sbornik: materialy dlia istorii razvitiia khimii v Rossii* (Moscow, 1901).

[8] Two articles, first given at speeches, can be singled out: V. I. Vernadskii, "O znachenii trudov M. V. Lomonosova v mineralogii i geologii"; and N. N. Beketov, "Istoriia khimicheskoi laboratorii pri Akademii nauk," in ibid., 1-34, and 1-5, respectively (nonconsecutive pagination in text). Vladimir

educated public requested that this ceremony not be restricted to merely paying obeisance to Lomonosov's past contributions to the propagation of science in Russia, but that it must also have "at the same time a practical meaning for us, as well as for the future of that science of which Lomonosov was our first representative more than a century and a half ago."[9] Markovnikov may have been claiming Lomonosov for chemistry in this instance, but this could just as easily apply to science and learning generally.

Markovnikov's summons is repeated in varying guises by the other speakers, and symbolizes the efforts by Russian chemists to more securely elevate their status not by extolling the past services of fellow scientists to the country but by re-emphasizing how important the support of chemistry was to the country's development. As Russian chemistry became more established, especially institutionally, in the later part of the nineteenth century, there was the inevitable introspection that accompanied professionalization.[10] Despite the centrality of Lomonosov to the chemistry profession's efforts, and whatever the effects of introspection on the knowledge of chemistry's past in Russia, it was not accompanied by a substantive reappraisal of the work conducted by Lomonosov that was otherwise so extolled at the Moscow meetings. Even so, it would seem that the Academy of Science's ensuing fascination with

Vernadskii, a geologist and chemist, was one of the most distinguished scientists of his day, and remains a revered figure in Russia. His essay is a thorough account of Lomonosov's primarily mineralogical work, and he displays a rare judiciousness in evaluating Lomonosov's attainments historically. Beketov's item is a fine, if attenuated, discussion of Lomonosov's efforts to establish a type of physical chemistry in Russia.

[9]   V. V. Markovnikov, "Polutorastoletie russkoi khimicheskoi laboratorii," in ibid., 3.

[10]   On the emergence of the history of chemistry in Russia, much can be gained from Sheptunova, *Istoriograficheskii analiz rabot po istorii khimii v Rossii*, 19-74; and Solov'ev, *Istoriia khimii v Rossii*. On the status of both chemistry and the chemist in nineteenth-century Europe, profitable is Knight and Kragh, *The Making of the Chemist in Europe, 1789-1914* (Cambridge: Cambridge University Press, 1998). Knight and Kragh's work is one of the few historical surveys of "European science" that contains a discussion of developments in Russia (see Brooks, "The Evolution of Chemistry in Russia").

Lomonosov came at a most opportune time for a re-evaluation of his scientific legacy.

During the 1911 Lomonosov Jubilee there was the predictable deluge of literature that generally accompanies such occasions.[11] Among the Academy of Science's ambitious plans for the events,[12] the more important for Lomonosov's legacy were plans to search relevant archives for overlooked papers and documents concerning his activities and to ready them for publication. This would entail extensive translation efforts. The Academy was also determined to bring to completion the latest and fullest version of Lomonosov's collected works (which had been launched under Mikhail Sukhomlinov's direction in 1891);[13] to sponsor specialized collections of articles focusing on Lomonosov's heterogeneous legacy; to compile bibliographies encompassing both his own writings and materials about him in Russian and in several foreign languages;[14] and to organize a special exhibition devoted to mid-eighteenth-century Russian culture, termed "Lomonosov and the Elizabethan Times."[15] Each of these efforts was eventually realized.

11   E. B. Ryss, "Bibliografiia osnovnoi literatury o M. V. Lomonosove za 1911-1916 gg.," in *Lomonosov: sbornik statei*, vol. 3, 587-606; "Lomonosovskiia torzhestva," in *Pamiati Lomonosova. Sbornik statei*, 88-105; and "Ukazatel' iubileinoi literatury o Lomonosove," in ibid., 106-22.

12   Kuliabko, "Lomonosovskii iubilei," 300-01; and Radovskii, *Lomonosov i Akademiia nauk*, 249-53.

13   Lomonosov, *Sochineniia*, volumes 1-5 (St. Petersburg, 1891-1902). Sukhomlinov died shortly before the fifth volume was released.

14   These bibliographies, which came out within a few years of the jubilee, were respectively Kuntsevich, *Bibliografiia izdanii sochinenii Lomonosova*, and Fomin, *Materaily po bibliografii o Lomonosove*. There were several essay compilations issued in and around 1911; perhaps the most rewarding remain Golubtsov, *Lomonosovskii sbornik*, which focuses on Lomonosov's connections to the far north of Russia, and *Lomonosovskii sbornik*, published by the Academy of Sciences in 1911, which offers several historical surveys of Lomonosov as a chemist and physicist.

15   *Putevoditel' po vystavke "Lomonosov i Elizavetinskoe vremia"* (St. Petersburg, 1912). An extensive showing of eighteenth-century cultural artifacts was held at the Academy of Arts in 1912. As is clear from the guide, the

The most grandiose undertaking suggested during the jubilee, however, which was for the establishment of a large-scale research institute devoted chiefly to chemistry, physics, and mineralogy which was to bear the moniker of the "founder" of said sciences in Russia, was never realized.[16] This was the result, it would appear, of both the overly ambitious designs of its planners and the government's lack of interest in offering sufficient financial support.

What made the 1911 jubilee most significant, however, was that at this time Lomonosov's most accomplished "modern" biographer, the historian and physical chemist Boris Menshutkin (1876-1938),[17] began to add a substantial gloss to representations of Lomonosov specifically as a chemist and physicist. Menshutkin was one of Russia's first historians of science, and unquestionably its most prolific early historian of chemistry. It appears that Menshutkin came by his interest in Russia's scientific past naturally, for his father was Nikolai Menshutkin, a noted chemist and also

---

exhibit was partially an attempt to closely associate Lomonosov with the government's seemingly long-term encouragement of Russian science and education.

[16]   On the proposed Lomonosov Institute, see A. V. Kol'tsov, "Proekty organizatsii Lomonosovskogo instituta v Akademii nauk v nachale XX v.," in Lomonosov: sbornik statei, vol. 6 (Moscow-Leningrad, 1965), 294-300. The institute was meant to recognize both the diversity of Lomonosov's interests and the apparent melding within his career of theory and practice. Although the author lays most of the blame for the failure of the Academy to establish the research center on the government, it would seem that a more likely final explanation was the onset of war in 1914—a factor that Kol'tsov downplays.

[17]   A. M. Smolegovskii and Iu. I. Solov'ev, Boris Nikolaevich Menshutkin: khimik i istorik nauki (Moscow, 1983), is a finely researched biography that includes meticulous coverage of Menshutkin's work as a chemist and less reliable attention to him as a historian of science (principally of chemistry). A useful chronicle of Menshutkin's interest in Lomonosov is S. A. Pogodin and N. M. Raskin, "B. N. Menshutkin kak issledovatel' trudov Lomonosova po khimii i fizike," in Lomonosov: sbornik statei, vol. 6, 245-66. In English, see also Tenny L. Davis's foreword to Boris N. Menshutkin, Russia's Lomonosov: Chemist, Courtier, Poet, trans. Jeanette Eyre Thal and Edward J. Webster (Princeton: Princeton University Press, 1952), v-viii.

a historian of chemistry.[18] In addition to his work on Lomonosov, Boris Menshutkin composed treatises on many leading chemists from the Russian past, including: Vasilii Petrov, Nikolai Zinin, Dmitrii Mendeleev, a large study of his father Nikolai Menshutkin,[19] and interestingly, a piece on Vasilii Severgin.[20] These works were preliminary to his planned general history of Russian chemistry, which he was not able to bring to fruition.

Menshutkin left a brief autobiography (penned in 1937)[21] that unfortunately has little value for those interested in, in lieu of a better expression, the psychological roots of Menshutkin's devotion to Lomonosov. It does, however, provide a framework within which to pinpoint the origins of his interests, or at any rate a sense of how he wished his first encounter with Lomonosov to be conveyed. Rather by chance, it seems, Menshutkin became aware of Lomonosov when as a student he attended a chemical society meeting (in 1900) and listened to one A. A. Zhivkov speak of Lomonosov's services as a chemist. According to Menshutkin, he was inspired to examine Lomonosov's place in the history of chemistry by this talk: "I attempted to track down any information

---

[18]   Sheptunova, *Istoriograficheskii analiz rabot po istorii khimii v Rossii*, 60-62.

[19]   Outside of his work on Lomonosov, Menshutkin's biography of his father, *Zhizn' i deiatel'nost' Nikolaia Aleksandrovicha Menshutkina*, St. Petersburg, 1908, is his most substantial publication in the history of science. It is also an account, from the perspective of a devoted son of course, of the older Menshutkin's efforts to improve university governance (he was long affiliated with St. Petersburg University), and his often contentious, though not oppositional, stance toward the regime.

[20]   Where he made an explicit effort to establish a link between Severgin's and Lomonosov's attempts to "disseminate enlightenment" in Russia, as cited in Smolegovskii and Solov'ev, *Menshutkin*, 130. For more on Menshutkin's various biographical efforts, see ibid., 120-50.

[21]   PFA RAN, f. 327, op. 1, no. 110, ll. 11-25ob. This was also published in Smolegovskii and Solov'ev, *Menshutkin*, 7-32. Menshutkin's memoir is largely annalistic in structure. Paradoxically, his "autobiography" is so utterly lacking in introspection that it bestows upon the few insights into his life that he does provide an air of authenticity (this despite the fearful year it emerged).

I could about him in the chemical literature, but I found nothing."[22] Evidently dismayed at this absence of materials, "I then decided to investigate the matter myself and commenced with the study of what documents, memorandums and notes were located in the manuscript division of the Library of the Academy of Sciences and in the archives." Following his earliest investigations,[23] Menshutkin would proceed to base the remainder of his nearly forty years of work on Lomonosov both on unearthing and bringing to light his actual papers, and far more consequentially on interpreting their enduring meanings for what was principally a non-specialist audience.

The element of chance, or of providence as it were, in Menshutkin's original "discovery" or "rediscovery" of Lomonosov has deep resonance and is itself central to the evolution of Lomonosov's image.[24] This difference between discovery and re-

---

[22]   PFA RAN, f. 327, op. 1, no. 110, l. 13; and Smolegovskii and Solov'ev, *Menshutkin*, 10-11.

[23]   Menshutkin rather quickly went on to compile *Lomonosov Considered as a Physical-Chemist: Toward a History of Chemistry in Russia* (*Lomonosov kak fiziko-khimik: k istorii khimii v Rossii*, St. Petersburg, 1904). In this volume, he included in whole or in part eighteen of Lomonosov's treatises in physics and chemistry, many of which he translated from Latin, and at least half of which had not been published before. He also added extensive commentary to the papers, speeches, and dissertations. For this entry, Menshutkin was awarded a prize (in the amount of 500 rubles) given by the Academy of Sciences. On this see Radovskii, *Lomonosov i Akademiia nauk*, 244-46. Created in 1868 ("O premii za uchenoe zhizneopisanie Lomonosova," *Zapiski Imperatorskoi Akademii nauk* 31, book 1 [1880]: 229-31) the premier citation, which came with 2000 rubles, had not yet been won: it was intended for a scholarly, and comprehensive, memoir. Although Menshutkin's work was not a full biography, it was so well received that the Academy deigned to grant him a reduced prize.

[24]   Pivotal to this idea of rediscovery is the fate of many of Lomonosov's apparently missing papers, which often reads, or rather has been written, like a mystery, and is extensively examined in Kuliabko and Beshenkovskii, *Sud'ba biblioteki i arkhiva Lomonosova*, 73-143. The notion of lost, or rather expropriated, papers was first implanted into the historiography by Lomonosov's earliest memoirists. Staehlin alleged that after Lomonosov's death "all of his manuscripts came into the possession of Count Grigorii Orlov" (who was Lomonosov's last patron of note), "Konspekt

discovery is not merely a semantic point. It concerns the more fundamental idea of how widely known Lomonosov was as a chemist and physicist, as opposed to his renown as a litterateur, before Menshutkin's researches. Although Menshutkin relied heavily on the documentary collections put together by Biliarskii, Kunik, and Pekarskii,[25] he sought original and incisive evaluations of Lomonosov's science, and for this, at least as he most often argued, the level of especially chemistry before his own time was insufficient to provide them.

Menshutkin's own work is clearly motivated by the notion of his personal discovery of Lomonosov, both the actual physical remains of his treatises and the scientist himself. This was an effort by him to signify his distinctive role in the rather expansive Lomonosov industry.[26] Most Soviet historians of Russian science,

---

pokhval'nogo slova Lomonosovu," 25. Also, soon after Lomonosov's death, his longtime nemesis at the Academy of Sciences, Johann Taubert, wrote (his dispatch is dated 8 April 1765) to Gerhard Müller that "on the day after his [Lomonosov's] death Count Orlov had his office sealed. Without a doubt there were located within it papers, which it was desirable not to have [allowed] released into someone else's hands." Pekarskii, *Dopolnitel'nye izvestiia*, 88-89. Documents of some value may have been taken from Lomonosov's study after his demise; there is, however, despite the resources spent on the study of him, no irrefutable evidence to support such a contention. Whatever the answer, the idea that Lomonosov was working on some potentially controversial work, not necessarily related at all to his scientific exercises, is, not surprisingly, a very evocative one in the historiography.

[25] Although Menshutkin nowhere singles it out for praise, perhaps most beneficial to him was the work of the philologist Anton Budilovich, whose 1869 work *Lomonosov kak naturalist i filolog* displays what was for the time an unmatched familiarity with Lomonosov's papers—both scientific and literary—which were housed in the Archive of the Academy of Sciences. Budilovich also excerpts at some length Lomonosov's chemical and physical dissertations; however, he rarely comments on them.

[26] Menshutkin is hardly above providing an intrepid sheen to his own toils as Lomonosov's biographer, as if forty years of labor does not provide enough evidence of his tenacity, if not necessarily valor. As he writes in his autobiography, despite the terrible privations that he and his mother underwent in Petrograd in 1919-20 during the Russian civil war, he still found the energy "to offer a course on organic chemistry to three students

desirous of emphasizing that Lomonosov was long celebrated in Russian culture, utilize the theme of rediscovery when discussing Menshutkin and Lomonosov[27]—that Lomonosov was widely admired both in his own time and by later generations of his admiring countrymen is axiomatic in arguably all accounts. It is the historians' subsequent speculations on his direct influence over later scientists that calls into question some of their contentions.

The less interesting aspect of the rediscovery trope is the apparent treasure trove of papers in the archives that Menshutkin uncovered and published, thereby establishing Lomonosov's farsightedness merely by their presence and Menshutkin's

---

and to labor on the history of chemistry—books on N. N. Zinin and M. V. Lomonosov." PFA RAN, f. 327, op. 1, no. 110, l. 16; Smolegovskii and Solov'ev, *Menshutkin*, 15.

[27] Sergei Vavilov perhaps most effectively, or influentially, attached this appellation to Menshutkin, maintaining that it was Menshutkin who "rediscovered" Lomonosov's pioneering status as a scientist, and specifically as a physical chemist. (Sergei Vavilov, *Mikhail Lomonosov*, 31). After Menshutkin, Vavilov is perhaps the most quoted modern source on Lomonosov's science. The rediscovery motif was, however, applied long before Vavilov, and made what was perhaps its initial appearance in the presidential address given by Alexander Smith to the American Chemical Society in 1911, when in the midst of an admiring review of Lomonosov's chemical research, Smith remarked apropos of the Russian scientist's reputation: "although his work in literary and linguistic lines, his success as a man of affairs, and his investigations as a geographer and a meterologist had won for him enduring fame, the fact that he was primarily a chemist had been completely forgotten. It was Menschutkin [sic] who, a few years ago, rediscovered him as a chemist, reprinted in Russian his scattered memoirs, and collected all that could be found of his manuscripts, letters, and laboratory note-books." Alexander Smith, "An Early Physical Chemist: M. W. Lomonossoff," *The Journal of the American Chemical Society* 34, no. 2 (February 1912): 112. Smith's essay was the first substantive study of Lomonosov's science to appear in English, and his final point that Menshutkin's "rediscovery of Lomonossoff [sic] has added at once a chemist of the first magnitude and a personality of marvelous force and range to the limited gallery of the World's very greatest men" (ibid., 119) has proven, unsurprisingly, to be warmly received by later enthusiasts of Menshutkin's achievements in the study of Lomonosov. On this see Pogodin and Raskin, "Menshutkin kak issledovatel' trudov Lomonosova," 260; and Smolegovskii and Solov'ev, *Menshutkin*, 115.

subsequent commentary. Far more engaging is Menshutkin's "recovery" of Lomonosov's importance as one of the most formidable scientific figures of the past two centuries worldwide, not only in Russia, and how within Russian scholarship he made the belief in this importance an article of faith that held strong for decades. If this seems to be less the disclosure of a real figure than the invention of an idealized one, the rediscovery metaphor is more apt.

Menshutkin's work on Lomonosov can be classified as follows: the unearthing and publication of hitherto unpublished, or seemingly forgotten previously published scientific treatises by Lomonosov; the accrual of extensive commentaries to said papers; specialized essays on Lomonosov's chemical and physics investigations; and the writing of more "popular" biographical studies. He published more than twenty (chiefly scientific) compositions covering with varying degrees of completeness every aspect of Lomonosov's natural philosophy.[28] Because this investigation attempts to unravel the more public mythology of Lomonosov, attention will be accorded exclusively to Menshutkin's popularization of his subject.

At the 8 November 1911 Lomonosov celebrations at the Academy of Sciences, Menshutkin delivered what was in retrospect the most striking, or at any rate historiographically eventful, speech of the event. Entitled *Lomonosov as a Natural Scientist* (*Lomonosov kak estestvoispytatel'*),[29] the speech introduces the paramount themes Menshutkin underscored throughout his nearly four decades of writing on Lomonosov. Moreover, it splendidly summarizes the full biography of Lomonosov that he issued that same year.[30] Due to the importance of jubilees to the creation and dissemination of Lomonosov imagery, and also because it situates representations of the scholar firmly in time and place, his speech is an exceptional

---

28   For a bibliography of Menshutkin's writings on Lomonosov, see ibid., 177-81.

29   Menshutkin, *Lomonosov kak estestvoispytatel'* (St. Petersburg, 1911; his discourse runs twelve pages).

30   Menshutkin, *Mikhailo Vasil'evich Lomonosov*.

window onto Lomonosov's depiction in the last years of the Russian Imperial era.

In trying to come to terms with Lomonosov's professional life, particularly its increasingly unfathomable diversity, Menshutkin alludes to Lomonosov's letter to Shuvalov in which Lomonosov ostensibly outlined his own preferences for the sciences over the other tasks to which he unwillingly bestowed so much time. Menshutkin, however, reformulated it to fit contemporary requirements:

> The activities of M. V. Lomonosov in the areas of Russian literature and philology already received in his lifetime wholly deserved appreciation, but until our times his name has been associated by almost everyone with that of a writer, one who created new forms of versification and who originated the modern Russian language. Meanwhile, Lomonosov mainly devoted his time to his work in his profession, chemistry and physics. However, his activities as a natural scientist have become well known in their entirety only in recent times.[31]

Menshutkin displayed this well-trodden point as an appeal to reevaluate the authentic nature of Lomonosov's importance in Russian history. He himself would never veer from its implicit demands that Lomonosov's science must receive further, indeed primary, exposure. Equally important is his stress, which he was the first to substantially develop, and which he made central to his approach towards Lomonosov, that it was only with recent developments in the sciences that Lomonosov's prescient research could be appraised from the proper perspective.

Like preceding biographers, Menshutkin allocates a considerable amount of time to alerting his listeners to the plainly astonishing details of Lomonosov's early biography.[32] Animated by the tales extolling Lomonosov's younger years, he deviated not at all from the myth. Of course, the educative purpose of portraying

---

[31]   Menshuktin, *Lomonosov kak estestvoispytatel'*, 1.

[32]   Ibid., 1-5.

Lomonosov in a manner that denoted amazement had lost none of its value. What is more, as a chemist, Menshutkin also clearly saw the importance of having such a stirring figure as Lomonosov as the progenitor of his profession. In reviewing Lomonosov's origins in the far northern periphery—and importantly, and stunningly, from the peasantry (albeit from the "enterprising" coastal dwellers of that region, the *pomors*); his hungry curiosity about nature, his love of learning, and his early and "passionate wish" to study the sciences, Menshutkin's reliance on earlier memoirs, particularly the eighteenth-century biographies of Lomonosov by Staehlin and Verevkin, for both "factual" information and their idealized narratives, is quite clear.

Lomonosov's work as a professor of chemistry and in establishing the first chemical laboratory in Russia, in addition to his more general tasks as an administrator and organizer of science, following his return from the "West" to St. Petersburg, are remarked on with deference.[33] Menshutkin outlined a few of the disparate non-scientific assignments that engaged Lomonosov, and "which constantly diverted him from his profession," such as "literary studies, work on history, philology and political economy," but given all these seemingly peripheral tasks, "it is in general amazing how much he was able to accomplish in the natural sciences."[34] These were, however, not mere trivial distractions, for "throughout his life Lomonosov always strove to bring the benefits of the pursuit of the enlightenment to the Russian people." His labors at the Academy of Sciences in popularizing science ("he was the first in Petersburg to give public lectures in physics"), his translation work and diverse published writings, his direction of the Academy's gymnasium and university, and finally his drive to establish Moscow University

---

[33]   Ibid., 5.

[34]   Ibid., 6. Lomonosov's literary, historical, and philological studies have received mention, and as for his work on "political economy," Menshutkin is most probably alluding to Lomonosov's paper (addressed in the form of an epistolary appeal to Ivan Shuvalov): *O sokhranenii i razmnozhenii rossiiskogo naroda* (*On the Preservation and Multiplication of the Russian People*, 1761), in Lomonosov, *PSS*, vol. 6, 381-403, 596-600.

all helped to impart to the Russian public a cognizance of the significance of science and learning.

None of this came without enormous struggles on Lomonosov's part, and Menshutkin was rather more engrossed than Lomonosov's previous biographers had been in trying to perceive and clarify the motivations behind his often-combative encounters with colleagues and contemporaries. For, Menshutkin observed, in order to get a fuller picture of the man, the "less pleasant side of his character" would also have to be illuminated.[35] Due to Lomonosov's meteoric rise from the geographic and social margins of Russian society, which was always rendered as exceptional, along with his apparently unrefined personality traits—the result no doubt of his lowly origins (again this was an established point that was employed at times to explain his temperament)—he developed a "high opinion of himself which compelled him to believe that his conclusion to every question was final and indisputable and that every objection was a personal attack."

From this conceit arose "endless battles" with others at the Academy who "he saw as hindrances to the diffusion of enlightenment in Russia, who appeared [to him] to be the persecutors of science." This perhaps antagonistic and uncompromising side of Lomonosov's character came with a high price, for these skirmishes, "which became especially common and sharp in his old age," and which along with his "incessant monetary woes," as well as his "predilection for indulging in spirits," gradually undermined him and led to the almost complete cessation of productive work by him during his last years. For Lomonosov the stark outcome of his own choleric disposition and of his alcoholism was that he died, as Menshutkin put it, still "relatively young," on 4 April 1765.

Earlier memoirists often hinted at Lomonosov's disagreeable temperament and his incessant battles with various enemies, both real and presumed; they were less explicit, however, about speculating on the effects of such behaviors on Lomonosov, and even less so regarding their effects on the Academy of Sciences.

---

35   Menshutkin, *Lomonosov kak estestvoispytatel'*, 7.

Menshutkin's apparent innovation here perhaps had less to do with any perceptual insights that he brought to bear in his studies of Lomonosov's character than with the evolution of the biographical genre itself.[36] Even as he exposed the less than admirable aspects of Lomonosov's life, however, they were still cast within his lecture as a whole in a heroic context.[37] Lomonosov was, after all, fighting to advance Russian science. So while Lomonosov seems to have been

---

[36]   At least as concerns Lomonosov, however, attempts to unravel his personality have gone little beyond Menshutkin's early forays. Given this lacuna in Lomonosov studies, E. P. Karpeev's "psychological portrait" of Lomonosov: "'Se chelovek...' (zametki k psikhologicheskomu portretu M. V. Lomonosova)," *Voprosy istorii estestvoznaniia i tekhniki*, no. 1 (1999): 106-21, a theme he also addresses in *Russkaia kul'tura i Lomonosov* (St. Petersburg, 2005), 9-25, can only be welcomed. Unfortunately, these compositions, the first that tackle this admittedly difficult subject, and from an able Lomonosov scholar, have none of the analytic sophistication that, for example, Frank Manuel decades ago brought to his examination of Newton (see Frank Manuel, *Isaac Newton*) or that John Banville effected in his vivid re-creation of Johannes Kepler, *Kepler: A Novel* (London, 1981). *Kepler*, part of Banville's "Revolutions Trilogy" (the other, in my view lesser, novels deal with Newton and Copernicus), exquisitely conveys Kepler's conflicted personality while also visualizing and contextualizing what it meant to be a natural philosopher during the so-called Scientific Revolution. Banville narrates an astonishing life, without losing a sense of the ideas, passions, and ambitions that drove Kepler forward or of the discoveries that we commonly construe as his legacy.

[37]   Menshutkin's portrayal of Lomonosov's turbulent life was candid, and though he often signaled some disapproval, he was not in the end condemnatory. Even over issues such as Lomonosov's slanderous (and drunken) behavior at Academy meetings, his failure to atone for which eventually led to him being put under house arrest (an incarceration that lasted from May 1743 to January 1744), Menshutkin could not bring himself to unambiguously censure him. In fact, he correctly emphasized that Lomonosov put the time of his arrest to great use in advancing his own studies. (This is only briefly dealt with in *Lomonosov kak estestvoispytatel'*, 5; for a somewhat fuller account, see his larger 1911 work, *Mikhailo Vasil'evich Lomonosov*, 30-35.) The period from 1742-44 was a chaotic time in Lomonosov's life, punctuated not only by his confinement but before that by a series of violent encounters with fellow employees at the Academy (chiefly with the Academy's "German" gardener). Documentation on these incidents is found in Pekarskii, *Istoriia Akademii nauk*, vol. 2, 329-48.

troubled with a temper that often detracted from what he might
have attained, did this not also give his real successes even more of
a miraculous aura?

Surveying Lomonosov's scientific work, Menshutkin was
convinced that there "could be little doubt that Lomonosov was
one of the outstanding chemists." Furthermore, "Lomonosov
accomplished enough in the areas of chemistry and physics,
wholly enough, for him to be called one of the greatest natural
scientists of the eighteenth century."[38] Menshutkin pronounced that
Lomonosov's essential bequest to succeeding generations of natural
philosophers was his innovative elaboration of the mechanical
philosophy (common to his times) to explain various topics, most
strikingly the nature of heat. Lomonosov's presumed anticipation
of the principle of the conservation of energy, along with similar
notions approximating a kinetic theory of gases, are indelibly
linked to his mechanical/corpuscular outlook on the makeup of the
natural world.[39] Menshutkin posited that these were revolutionary
hypotheses, far surpassing anything Lomonosov's contemporaries
had proposed, and remain of enormous relevance today. If that were
so, of course, then Menshutkin's inferences are perfectly logical, and
Lomonosov was a pioneering theorist.

Lomonosov was able to accomplish such extraordinary
advances in delving into the nature of heat and gas due to his
appreciation of the need for chemists to utilize physics and
mathematics in their work, and in utilizing them, his efforts
epitomize "the methods of the nineteenth century, and not the
eighteenth, when they were still not employed."[40] His application of
the techniques of these exact sciences to chemistry was an "entirely
original and independent point of view."[41] Moreover, the unity of
physics and chemistry achieved by Lomonosov stamped him as

---

[38]   Menshutkin, *Lomonosov kak estestvoispytatel'*, 9.

[39]   Ibid., 8-9.

[40]   Ibid., 11-12.

[41]   Ibid., 9, 12.

a physical chemist, of course Russia's first.[42] Recent developments in the maturation of chemistry, and especially of physical chemistry, would have made this point self-evident to Menshutkin's listeners. Therefore Lomonosov's work, in terms of both the methods he made use of and the propositions he formulated, was a precursor to contemporary research. Chemistry was the science for which Menshutkin most forcefully appropriated Lomonosov.

Menshutkin closed his oration with a consideration of the 1865 jubilee's significance in evaluating Lomonosov as a scientist. Ostensibly aimed at the rather weak previous understandings of Lomonosov, this passage in fact was mainly important in revealing where Menshutkin would endeavor to direct Lomonosov's renown:

> In 1865, when a century had passed since his death, in ceremonial gatherings of the Academy and University, scholars of the time issued evaluations of his works. In these speeches we find little indication of what today we would put down as most important in Lomonosov's works, such as his mechanical theory of heat and of gases, and physical chemistry; that these conceptions were not considered in 1865 is especially conspicuous; although a hundred years had passed since his death, and completely analogous physical theories were, prior to that time, already proposed by famous scientists of the nineteenth century, they were not disseminated widely in those days, and several more years were necessary before they gained admission into scientific use. The flowering of physical chemistry belongs only to the end of the past century, and these facts demonstrate how much of a genius Lomonosov was as marked by his times.[43]

Throughout Menshutkin's long years memorializing Lomonosov, his most defined aim was to attach a more modernized set of

---

[42]   Vladimir Markovnikov was perhaps the earliest scientist of note to contend that "Lomonosov was the first Russian physical chemist." See "Vstupitel'noe slovo pri otkrytii pervago zasedaniia zasluzhen. Prof. V. Markovnikova," in *Lomonosovskii sbornik: materaily dlia istorii razvitiia khimii v Rossii*, 15. He did not aver, however, as Menshutkin did, that Lomonosov was the first physical chemist, period.

[43]   Menshutkin, *Lomonosov kak estestvoispytatel'*, 12.

scientific signifiers to Lomonosov's biography. Menshukin's assertions concerning Lomonosov's theoretical acumen are difficult, perhaps even to a point unnecessary, to refute absolutely, for they are posed in such a general manner as to leave themselves open to virtually limitless interpretation.[44] Thus it is not the correctness of Menshutkin's assertions that will be subjected to scrutiny, but rather their effects on representations of their subject.

Mikhail Vasil'evich Lomonosov: a Biography (*Mikhail Vasil'evich Lomonosov: zhizneopisane*, 1911), along with its later slightly revised, or better said, expanded, editions, became the most consequential "large" memoir of Lomonosov' life that had yet appeared. Undertaking the project at the request of the commission organizing the 1911 Lomonosov Jubilee, Menshutkin wrote it with a lay audience in mind.[45] All of his works written prior to and during 1911 are re-

---

[44]   To my mind the best introduction to mechanical/corpuscular theorizing in the seventeenth and eighteenth century remains Boas, "The Establishment of the Mechanical Philosophy," 412-541. Boas's is one of the rare "western" studies that include a discussion, however brief, of Lomonosov. Excluding her assessment of Lomonosov's Newtonian affinity (a result, it would seem, of her reading of Menshutkin), her judgment is keen. For as she concisely concludes, by Lomonosov's time, or soon thereafter, "other systems [specifically Dalton's at the end of the eighteenth century] were less concerned with mechanical explanations and more with the characteristics of the atoms themselves" (p. 523). A disputatious, and unpersuasive, response to Boas is Lius Lanzheven, "M. V. Lomonosov i R. Boil' (korpuskuliarnaia teoriia materii i mekhanisticheskaia kontseptsiia mira)," in *Lomonosov: sbornik statei*, vol. 7, 49, 55-57.

[45]   Informative as a summary of Menshutkin's aims not only for this biography but in regard to Lomonosov generally was the plan he submitted for its composition to the Academy of Sciences in 1910. He was mainly interested in producing a work in "easily accessible language" that would meet an upsurge of interest in the study of the roots of Russian science. Lomonosov was the pivot around which this evaluation of the Russian scientific past would take place, for "many views and thoughts of Lomonosov which were expressed by him in his dissertations and scientific investigations have nowadays become commonly accepted and are not seen, as they were in his time, as strange and incomprehensible." He would also deal with questions of Lomonosov's "character and way of life" that would, presumably, along with a proper elucidation of his foremost achievements as a scientist, induce considerable interest in Lomonosov's biography amongst the public.

flected in this biography.[46] It has usually been referred to as the most popular book of its type ("scientific" or "academic") in Russia up to that point.[47] Whether or not this was so depends on a rather loose analysis of both its press run and of its "type." Nonetheless, it has consistently been cast as such. Employed here will be the 1937 edition of the aforementioned memoir.[48] Except for the rare inclusion by Menshutkin of Soviet-inspired, or Soviet-necessitated, rhetoric,[49]

As cited in Pogodin and Raskin, "Menshutkin kak issledovatel' trudov Lomonosova," 258-59.

[46] Menshutkin's involvement with studying Lomonosov was perhaps most intensive in 1911. In addition to an active role on the commission planning the jubilee (he was added to the Academy of Science's organizing committee soon after its formation), he had also been increasingly immersed since 1907 in efforts led by the aging philologist and long-time student of Lomonosov, V. I. Lamanskii, to conclude two further "science" volumes for the long-delayed completion of Lomonosov's complete works. They were to come out by 1911, but due initially to editorial problems and later to tumultuous conditions within Russia and later the Soviet Union they were only issued in 1934. On the assembly of these two volumes, see Menshutkin's preface to Lomonosov, *Sochineniia*, vol. 6, 1934, V-IX. Furthermore, Menshutkin contributed a pair of articles: "O korpuskuliarnoi filosofii Lomonosova"; and "M. V. Lomonosov i flogiston," for *Lomonosovskii sbornik* (St. Petersburg, 1911), 151-62, both of which are largely reprinted in his Lomonosov biography of the same year. Also that year he published "Trudy M. V. Lomonosova po fizike i khimii, "in *Trudy Lomonosova v oblasti estestvenno-istoricheskikh nauk* (St. Petersburg, 1911), 1-103. Here in whole or in part are found translations of several of Lomonosov's "physical-chemical" writings, all of them commented on by Menshutkin.

[47] It came out in an eventual press run of 80,000 copies. See Pogodin and Raskin, "Menshutkin kak isselodovatel' trudov Lomonosova," 259; Radovskii, *Lomonosov i Peterburgskaia Akademiia nauk*, 256; and Smolegovskii and Solov'ev, *Menshutkin*, 102. Menshutkin himself accented the popularity of the book in his autobiography: PFA RAN, f. 327, op. 1, no. 110, l. 23ob; Smolegovskii and Solov'ev, *Menshutkin*, 29.

[48] Menshutkin, *Zhizneopisanie Mikhaila Lomonosova*. Let it be noted that this edition is the one that was translated into English in 1952 under the title *Russia's Lomonosov*, and which is unequivocally the principal source for Lomonosov's life outside Russia. While all subsequent translations are my own, I have compared my efforts with the above English version.

[49] Such as the obligatory citation found in his preface noting the 1936 celebrations commemorating the two hundred and twentieth-fifth

this work merely amplifies without altering the Lomonosov that Menshutkin had developed in his 1911 study. Whatever minor distinctions do in fact exist between the texts have less to do with Stalinist political exigencies forcing him to revise the foundations of his earlier work than with simply an augmentation of detail—and they do not, it must be stressed again, include any substantial modifications in the main arguments or conclusions. The use of the 1937 edition also supports, implicitly, my judgment that the myth of Lomonosov was not substantially affected in content by the emergence of Soviet power. The consequences on the effects of the mythology of Lomonosov of excess exposure during the Soviet era are, however, quite another matter.

Menshutkin's *Lomonosov* is a complete biography, including the requisite retelling of the stories of Lomonosov's idealized youth and education that were fundamental to all representations of him since soon after his death. His later labors at the Academy of Sciences and in all the myriad scientific and non-scientific fields that were outlined by previous memoirists are given the lavish attention required in what was, after all, still in the main a hagiography. Juxtapositions between Lomonosov and Franklin (in the area of electrical research), and Lomonosov and Newton (in the area of optical research), are given perhaps more attention (and scientific polish) than found elsewhere earlier. Other dimensions

---

anniversary of Lomonosov's birth. Here he makes mention of *Pravda's* headline article (18 November 1936) on Lomonosov, which hailed the "Brilliant Son of the Great Russian People." Partially taking his cue from *Pravda's* nods towards Lomonosov's value as a symbol to "Soviet youth," he intones: "The life and activity of Lomonosov, the great patriot, the genius scientist, the passionate fighter for an original science and culture, are very instructive in our era, particularly for the coming generation." Menshutkin, *Zhizneopisanie Mikhaila Lomonosova*, 3-5. Replace Soviet with Russian, and like language is found in nearly two centuries of previous writings about Lomonosov. The 18 November 1936 issue of *Pravda* is primarily dedicated to (crudely) eulogizing Lomonosov's contributions to Russian science and culture and to acclaiming his lifelong struggles against the enemies of Russian advancement, with Lomonosov and Russia (and/or the Soviet Union), not surprisingly, conflated into a single representation.

indispensable to Lomonosov's constructed life, such as his founding of Moscow University, are also awarded lavish praise.

At first glance, what is particularly striking in Menshutkin's work is his attempt to seek coherence in Lomonosov's scientific activities—or perhaps it would be better to see it as his attempt to force the bewildering diversity of Lomonosov's professional life into a more clearly delineated whole. Since he was mainly interested in Lomonosov's legacy as a chemist—and to a lesser extent his legacy as a physicist—he primarily sought to establish that his chemistry and physics were conceptually subsumed under a rather accessible rubric. Allied with what would turn out to be an approachable and unified body of work was an unbounded heritage, which was also, paradoxically, rather simply defined.

For the purpose of organization, Menshutkin supplied a relatively porous chronological division which divided Lomonosov's work at the Academy of Sciences into physics (1741-48), chemistry (1748-57), and finally, and most amorphously, "applied sciences" and administrative spheres (1757-65).[50] This schematization, which has maintained its hold over later writers, is less relevant than his straining to aggregate Lomonosov's chemical and physical researches into a theoretically combined body of knowledge. This effort to demonstrate congruity was, even if restricted to Lomonosov's chemical and physical treatises, vital, for the very encyclopedic nature of Lomonosov's professional activities made them increasingly difficult to evaluate, particularly if one was interested in reinventing a life.

After inspecting Lomonosov's writings, Menshutkin discerned that he had attempted to blend "his scientific writings, especially those in physics and chemistry, into one well-ordered whole."[51] Physics and chemistry did not delimit the range of Lomonosov's science; Menshutkin insisted that mathematics was also intrinsic to it, and that science was not distinguished by an indecipherable

50   Menshutkin, *Zhizneopisanie Mikhaila Lomonosova*, 68.

51   Ibid.

heterogeneity, but rather that there was a determined purpose behind his research:

> From the beginning Lomonosov intended to write a great composition that would combine all of the aforementioned sciences on the basis of the corpuscular theory. Several times during the course of his life he strove to write such a "corpuscular philosophy" (as he termed it in one of his letters to L. Euler), but always some reason or other compelled him to stop at the very beginning, before he was barely able to outline a plan of the work. However, the different chapters of this great undertaking are almost all before us: those dissertations, speeches, and meditations which he communicated to the public, mainly at formal meetings of the academy.[52]

It was incumbent on Menshutkin not only to try to convey the array of valuable ideas to be found in the dozens of disparate dissertations, many of them unfinished, which would be aided by classifying them all within the rhetorical device of a presumed comprehensive theory, but also to explain why these ideas had not been properly recognized either at home or abroad. Whether or not Lomonosov in fact planned to write a work combining his ideas is, given the vagueness of his references to it, such as in the letter to Euler,[53] rather difficult to ascertain. Menshutkin introduces the elements that would frame his attempts to build an authoritative life of Lomonosov: intellectual unity, a recognized authority (in the familiar guise of Euler) able to endorse Lomonosov's worth, and the (re)discovery of his principal contributions to science.

In addition to formulating a grand atomic/corpuscular theory unifying Lomonosov's theoretical efforts, it was presenting him as a physical chemist, the prototype for the profession, and an individual personifying the merger of physics, mathematics, and chemistry into one, that was now deemed most valuable in developing his scientific legacy. Menshutkin does not uncritically

---

[52]   Ibid., 67.

[53]   Lomonosov *PSS*, vol. 10, 450-51, 57.

accept Lomonosov's science as being inviolate, but even when imposing restrictions on its practical import he allows for it to possess a startlingly rich potential.

In concentrating, correctly, on Lomonosov's unreservedly mechanical or corpuscular explanations for natural phenomena as the theoretical approach under which nearly all of his chemical and physical writings can at least roughly be subsumed, Menshutkin describes its fundamental proposition as follows:

> Lomonosov's principle is the chemical element, as it was characterized by Robert Boyle in 1661: a simple body incapable of being additionally broken down by means of chemical analysis. Little by little in the eighteenth century this conception found favor among chemists until after several decades it was made the basis of Lavoisier's doctrine of chemical elements. It is extremely interesting what Lomonosov further conveys about "elements" and "corpuscles": elements are in essence the atoms of the chemists, and corpuscles—the molecules. We have here the first combination, the first unification of two conceptions of the elements, which takes its beginnings from extreme antiquity: the first talks of elements as qualities, and according to the second, the elements are atoms—these are the smallest further indivisible primary particles of all bodies.
>
> The unification of these two points of view was brought forward by Lomonosov, introducing as the main proposition an understanding of the corpuscle-molecule as having exactly the same quantitative composition as the corresponding body it forms.[54]

So it was not simply a crudely offered anticipation of later ideas that Menshutkin offered—he was too careful a historian of science for that. Rather, he situated Lomonosov's corpuscular views within an impressive genealogy of atomic thinking, and more compellingly in a direct line between the conceptualizations of Boyle and Lavoisier. The Boyle association is important not only in that it situates Lomonosov's apparently equivalent hypothesizing, but also in that it emphasizes Lomonosov's own education and the

---

[54]   Menshutkin, *Zhizneopisanie Mikhaila Lomonosova*, 142.

probable influences on him.[55] As for Lavoisier, this is yet another link between Lomonosov and more traditionally recognized scientific figures.

Underlying Menshutkin's presentation is the question of why Boyle, Lavoisier, and indeed a host of others whose insights

---

[55]   Robert Boyle's natural philosophy was the greatest influence on Lomonosov's scientific speculations—especially in chemistry and physics. Lomonosov referred in a substantive manner to Boyle's work more often than he did to any other natural philosopher of the time. Boyle's prestige had, however, by the early eighteenth century been utterly eclipsed by that of Newton. Euler and Wolff were, clearly, more valuable signifiers in the St. Petersburg Academy. Lomonosov's profound intellectual debt to Boyle is best verified by a reading of his corpuscular papers (see in particular Lomonosov's *Meditations on the Cause of Heat and Cold* and *Physical Meditations on the Cause of Heat and Cold*, in Lomonosov, *Polnoe sobranie sochinenii*, vol. 2 [Moscow-Leningrad, 1951], 7-55, 63-103, 647-53). In fact, his mechanical perspective was drawn in large measure from Boyle's ideas, however much he may have differed with Boyle in drawing certain inferences, such as those concerning the nature of fire, or rather the existence of a caloric material, in his theorizing. Henry Leicester's "Boyle, Lomonosov, Lavoisier, and the Corpuscular Theory of Matter," and idem, *Lomonosov on the Corpuscular Theory*, 13-46, passim, explore Lomonosov's reliance on Boyle. For Lomonosov's familiarity with the breadth of Boyle's writings, consult, albeit with a degree of skepticism, Korovin's *Biblioteka Lomonosova*, 92-101. Among the works of Boyle drawn on by Lomonosov are: *Certain physiological essays and other tracts, written at distant times and on several occasions...The second edition, wherein some of the tracts are enlarged by experiments, and the work is increased by the addition of a discourse about the absolute rest in bodies* (1669; he used the 1677 Latin edition); *Essays on the strange subtility, great efficacy and determinate nature of effluviums...*(1673; he used the 1677 Latin edition); *Historia fluiditatis et firmitatis* (1667 and 1677); *New experiments physico-mechanical touching the spring of the air and its effects (made for the most part in a new pneumatical engine* (1660; he used the 1661 Latin edition); *A continuation of new experiments physico-mechanical touching the spring and weight of the air and their effects. The 1-[2] part...* (1669-1682; he used the 1682 and 1685 Latin editions); *The origins of formes and qualities (according to the corpuscular philosophy) illustrated by considerations and experiments, plus The second edition, augmented by discourse of subordinate frames* (1666-1667; he used the 1671 and 1688 Latin editions). Boyle's articulation of a corpuscular conception of nature, along with an elucidation of his place in the history of atomic philosophizing, is scrutinized in William R. Newman, *Atoms and Alchemy: Chymistry & The Experimental Origins of the Scientific Revolution* (Chicago, 2006); and Boas, "The Establishment of the Mechanical Philosophy."

were not, in theory, perceptibly more advanced than Lomonosov's received the entirety of the renown. There is the compelling need in Menshutkin's arguments to explain why Lomonosov's notions, which were precursors to later advances, and were now acknowledged, at least by Lomonosov's more uncritical admirers, as basic to explaining the division of matter, went unrecognized. One answer might be found in the rather less developed techniques of Lomonosov's time, or, as characteristically reasoned by Menshutkin:

> Lomonosov's theory is close to that of Dalton, who called a corpuscle or a molecule of a complex body a complex atom. But, as a predecessor of Dalton's, Lomonosov did not have those precise quantitative facts which Dalton already possessed, and which were the result of the development of chemical quantitative analysis in the last quarter of the eighteenth century. And without those quantitative facts it was inconceivable to elaborate a chemical atomic theory: for only those facts gave it the necessary bearing.[56]

This is the crux of Menshutkin's analysis, for even as he points out the brilliance of Lomonosov's ideas he sees their limitations given the age he lived in. While this might explain why Dalton, for example, received honors, it neglects to offer a rationale for why both Boyle and Lavoisier did as well. So the answer, a well-rehearsed one in the study of Lomonosov, is that the fault lies not in any possible absence of discernment on Lomonosov's part that prevented his work from being appropriately received, but in the less developed state of chemistry in the eighteenth century. This, of course, still begs the question of why Lomonosov's hypotheses were slighted while those proposed by many of his less deserving contemporaries found support.

At other times, however, when trying to account for Lomonosov's seeming obscurity, particularly in not having his atomic/corpuscular theorizing acclaimed, Menshutkin maintained that Lomonosov's "writings played no role" in contemporary

---

56   Menshutkin, *Zhizneopisane Mikhaila Lomonosova*, 142.

scientific debates since most of his relevant treatises "remained unpublished during his lifetime."[57] He also noted, in a remark that in like form was scattered throughout the biography, that they "were first published in my translation in 1904." Although it is not vital to belabor a point dealt with earlier, Lomonosov's most important corpuscular paper, *Meditations on the Cause of Heat and Cold*, was published in *Novi Commentarii* (1750) and received fairly wide, albeit highly critical, attention at the time.[58]

The reasons for the tension in Menshutkin's discussion are readily observable. On the one hand, he argues that the failure to adequately acknowledge Lomonosov's services to Russia was primarily due to the underdeveloped nature of the sciences of the time, which left little room for the prescient researcher to be accepted. What this surely means is that the significance of Lomonosov's work or achievements could not be appreciated until chemistry and physics had matured to the level where his papers would be understood. On the other hand, Menshutkin advances the idea that the more singular cause of Lomonosov's near anonymity was the failure of his treatises to receive either suitable exposure in print, or their having been left unpublished and forgotten in the archives. This in turn should, it would seem, inspire or stimulate the search for even more of his surviving papers by later scholars. Yet there is an unfortunate fact plaguing Menshutkin's reasoning: Lomonosov's more important corpuscular dissertations were indeed published, in Europe. Undaunted by such inconsequential impediments to his fashioning of Lomonosov, Menshutkin's resolution is that if Lomonosov's corpuscular viewpoint "would have been published in connection with all its later developments, it might, perhaps, have had a considerable meaning for the cultivation of physics and chemistry."[59] But it was not properly disseminated, so it was up to Menshutkin to illuminate Lomonosov's legacy at a time when Dalton and his successors had made its value purely academic.

---

[57]　　Ibid., 76.

[58]　　Pavlova, *Lomonosov v vospominaniiakh*, 151-58.

[59]　　Menshutkin, *Zhizneopisanie Mikhaila Lomonosova*, 76.

Critical also to Menshutkin's approach is the concept of quantitative methods, and this is one to which he repeatedly returns. Lomonosov's memoirists since the eighteenth century had been aware of the need to apply a mathematical referent to his biography, for if his methods were rational, or "correct," then his worth as a symbol to later generations would be even greater. Lomonosov's mechanical outlook on nature permitted very liberal readings of its probable influence, as well as easy dismissals of it, but if he also had the accompanying analytical skills, then his corpuscular theory would be even more esteemed by later chemists and physicists.

According to Menshutkin, Lomonosov became aware of the need to supply his work with mathematical proofs from Christian Wolff.[60] Well aware of the weaknesses marking Wolff's own employment of mathematical analysis, Menshutkin eschewed any rigorous discussion of mathematics itself; instead he credited Wolff's methodology, or rather his "mathematical philosophy," with deeply influencing the form of Lomonosov's arguments, for it permitted Lomonosov "to develop and express his original thoughts in a strict logical sequence." This is not to say that Lomonosov's natural philosophy was close to Wolff's, for Menshutkin argued forcefully that, despite some superficial similarities, it was not.[61] However, a vague stress on Wolff's mathematical methods, even if they has little do with the application of analysis to natural phenomena, was what he clearly wanted to impart to the reader.

Lomonosov's *Oration on the Usefulness of Chemistry* (1751), one of his most quoted pieces, is excerpted at considerable length in Menshutkin's work. What Lomonosov says of mathematics became a precious resource to later scholars, even though it reveals little more about the topic than did the Wolff reference. "'Useless are eyes for those who wish to see the interior of a thing, but lack hands to open it,'" writes Lomonosov, while "'useless are hands for those who have no eyes to examine the things that have been revealed. Justly Chemistry can be called the hands, and Mathematics the

60   Ibid., 75.

61   Ibid., 42, 75, 77.

eyes of Physics.'"[62] But chemistry and mathematics are as yet estranged, for the chemist disdains the mathematician '"as one who is practicing only some futile reflections about points and lines'"; while the mathematician disregards the chemist for '"being preoccupied solely with practice and ... lost among many disorderly experiments.'" This alienation is to the detriment especially of chemistry, for as opposed to physics, which is inseparable from mathematics, '"chemistry had yet to be joined with a thorough knowledge of Mathematics,'" and until it was, it would be unable to supply the necessary experimental proofs so vital to its further development as a science.

Lomonosov's brief comments offer a wonderful introduction not merely to his apparent awareness of the long-term significance of mathematical analysis, but also to the integration by him, or rather by Menshutkin, of chemistry, physics, and mathematics into the field of physical chemistry. Much time is spent detailing Lomonosov's efforts to establish a chemical laboratory, including the cumbersome preparation, indeed the invention, of an array of laboratory equipment (the designs for which Menshutkin uncovered), the training of students, and the preparation of general courses and lectures on physical chemistry.[63] In short, as read here, Lomonosov laid the foundations for the study of chemistry in Russia. Given that he in fact spent perhaps less than a year offering lectures in chemistry, and except for the prematurely deceased Klement'ev left no "school" behind, it's evident that Menshutkin relies heavily in this discussion on a highly speculative construal of what the potential for the chemical laboratory may have been, rather than on what Lomonosov actually accomplished there.[64]

Menshutkin utilized Lomonosov's *Oration on the Usefulness of Chemistry* to outline how defined physical chemistry. That

---

[62]   Ibid., 144-45.

[63]   Ibid., 150-61.

[64]   For an even more generous rendering of the chemical laboratory's—and Lomonosov's—shaping of both contemporary and later Russian chemistry, see Raskin, *Khimicheskaia laboratoriia Lomonosova*.

this speech was so repeatedly drawn on by Menshutkin again underlines that it was not Lomonosov's actual unearthed papers that distinguished Menshutkin's shaping of him, but rather the motif of their long-hidden import. As for physical chemistry, as Lomonosov proclaimed it:

> Physical chemistry is a science, which explains on the foundations of the theses and experiments of physics the reason for what occurs through chemical operations in complex bodies. It perhaps may be called chemical philosophy, but in an absolutely different sense than that mystic philosophy, where not only are the explanations not given, but even the operations themselves are conducted in secrecy.[65]

This wonderfully vague extract provides another reinforcement of the notion of method over content. It stresses the need to rigorously study those particles central to Lomonosov's corpuscular views, and to do so on an implicitly mathematical basis, which would raise the standard of chemistry, if in a typically imprecise direction. As proffered by Menshutkin, whatever the substance of Lomonosov's chemical dissertations, his was a rationalized approach to chemistry with the aid of physics. This fits perfectly with the educative aspect of Lomonosov as a chemist. Worth noting is that the elimination of so-called occult forces was doctrinal to "mechanically minded" seventeenth- and eighteenth-century natural philosophers.

Physical chemistry is the discipline with which Menshutkin most associated Lomonosov, to the point that its origins became indistinguishable from Lomonosov in his narrative. But even so,

---

[65]   Menshutkin, *Zhizneopisanie Mikhaila Lomonosova*, 155. As pointed out by Henry Leicester, it is important to keep in mind that "physical chemistry," a term Lomonosov often used to describe his laboratory work, was applied by him because "he felt that the theoretical, or philosophical, side of chemistry required a rigorous treatment if chemistry was to become a true science.... It would be well to recognize, however, that, as Lomonosov himself said, the term to him meant the same thing as 'chemical philosophy', that is, theoretical as opposed to practical chemistry and not what the modern chemist means by this expression." Leicester, *Lomonosov on the Corpuscular Theory*, 18-19.

chemistry had not yet reached the requisite stage—and this element, again, is underscored throughout Menshutkin's analysis—to allow Lomonosov's "physic-chemical experimentation" to offer up a satisfying series of proofs. Lomonosov had to devise not only the chemical equipment itself, but also even the techniques of analysis. Given these severe limitations, Menshutkin had to plaintively admit, as regards his legacy as a physical chemist,

> that here, as in the other areas of Lomonosov's scientific work, we have very valuable thoughts, and a brilliant foreknowledge of those roads on which the further development of science must progress; however, of practical accomplishments from these thoughts and intentions there were no results due to the utter absence of instruments, devices, and methods of investigation. The ideas outstrip the practical resources by a century and a half.[66]

So, despite the trailblazing quality of Lomonosov's efforts as a physical chemist, no significant or even measurable consequences came to pass. However, after more than a century had elapsed, in the 1880s to be more precise, the beginnings of physical chemistry charted with such promise by Lomonosov were taken up again when Wilhelm Ostwald ("also Russian by origin" Menshutkin informs us) became "one of the first and most important figures in this trend."[67] He then goes on to reveal that he himself informed Ostwald of Lomonosov's earlier work in 1905.

Despite the lack of any immediate intellectual response to Lomonosov's "chemical philosophy," in the following decades, as chemistry became more quantitatively (or mathematically) based,

> that trinity that was at one time proclaimed by Lomonosov, of chemistry, physics, and mathematics, has become an

---

[66]   Menshutkin, *Zhizneopisanie Mikhaila Lomonosova*, 158. It was in *Ostwald's Klassiker der exakten Wissenschaften* (1910, no. 178), that several of Lomonosov's chemical and physical papers, translated into German in part by Menshutkin, first found their way before a non-Russian audience (excepting, of course, for some of their original appearances, in Latin, in *Novi Commentarii*).

[67]   Menshutkin, *Zhizneopisanie Mikhaila Lomonosova*, 160.

accomplished fact. Weight, measurement, and number were introduced into chemistry with the assistance of physics, transforming it into an exact science, so now too chemistry has begun to penetrate ever more into physics, and in doing so is forming a chemical physics. Both sciences are unthinkable without the other, and also without mathematics, as was clearly seen by Lomonosov; both supplement one another, and contribute towards mutual conquests in the sphere of the unknown. He was the first physical-chemist, the father of physical chemistry.[68]

Physical chemistry most persuasively reinforces Menshutkin's notion that Lomonosov's work was conducted in a unified and rationalized manner, a manner that could be duplicated once that science had been revived. Therefore, although Lomonosov's lack of contemporary recognition was inevitable given that as a physical chemist he was unique in his age, over the long term (which for science is the more important measure) his achievements could not be disputed. Manifest also in Menshutkin's rendering is his striving to establish a consequential link between Lomonosov and later physical chemists; a connection perhaps occasionally difficult to detect, but one that nonetheless eventually encouraged generations of (unnamed) Russian scientists to follow in Lomonosov's path.

An aspect of Lomonosov's work as a physical chemist that has had wide resonance in the literature concerns his apparent anticipation of the law of the conservation of matter. Menshutkin too did not deny Lavoisier's fundamental role in offering hypotheses that would decades later coalesce into an apparent "law." He was quite intent, however, on demonstrating that Lomonosov was working in similar areas, and came up with ideas that, however tentatively posed, indicated that he shared credit with Lavoisier for helping to usher in a "revolution" (my term) in chemical thinking.[69] To corroborate his reasoning, Menshutkin mainly utilized a slender series of remarks that Lomonosov first announced to Euler in 1748,

---

68   Ibid.

69   Ibid., 145-50.

and which were repeated almost verbatim by him in a dissertation delivered at the Academy of Sciences in 1760.[70] Menshutkin also referred to a sequence of experiments that Lomonosov conducted, although there is little evidence either to verify that any substantive experimentation actually took place, or to tie the miniscule amount perhaps accomplished to any of the later research undertaken by Lavoisier.

What does exist is an oft-repeated phrase of Lomonosov's in which he makes even-then distinctly obvious observations:

> All of the changes that occur in Nature conform to the law that as much is taken away from one body, so much is added to another. In other words, if matter is reduced in one place, it must increase in another place; [additionally] as much time as one gives over to being awake, the same amount of time is deducted from sleep. This general natural law extends to the very rules of motion: for a body, which moves the other by its force, loses the same amount of force as it imparts to the other, which accepts motion from it.[71]

Philip Pomper subjected this statement of Lomonosov's, and more generally the suggestion of Lomonosov in some sense having anticipated Lavoisier, to a searching examination, and not surprisingly, given that said arguments largely rest on the above passage, found them without merit.[72] Menshutkin, as opposed to many later historians of science,[73] never claimed

---

[70]   For the full text of the Euler letter, see Lomonosov, *Sochineniia*, vol. 8, 1948, 72-91, (2) 18-22; idem, *PSS*, vol. 10, 439-58, 801. The speech, entitled *Rassuzhdenie o tverdosti i zhidkosti tel* (*A Dissertation on the Solidity and Liquidity of Bodies*), is found in Lomonosov, *PSS*, vol 3, 377-409, 559-65.

[71]   Menshutkin, *Zhizneopisanie Mikhaila Lomonosova*, 146 (Menshutkin is quoting from Lomonosov's above noted 1760 Academy address, for which see Lomonosov, *PSS*, vol. 3, 383).

[72]   Pomper, "Lomonosov and the Discovery of the Law of the Conservation of Matter," 119-127.

[73]   It would be extremely difficult to find a work on Lomonosov or on the history of Russian science published in the Soviet Union that does not in some fashion offer an argument supporting Lomonosov's priority over Lavoisier in at least visualizing the "law of the conservation of matter."

that Lomonosov was a direct precursor of Lavoisier; instead, he
offered that they had worked in similar areas, and came to broadly
comparable conclusions. Even such a vague contention is open

Although inspired by Menshutkin's speculations, the model for the more
extreme professions is perhaps Sergei Vavilov's "Zakon Lomonosova"
(*Pravda*, 5 January 1949). Given the cultural xenophobia that marked
the post-war years in the Soviet Union, during which magnifying (and
inventing) Russian and Soviet achievements was commonplace, with as was
an accompanying denigration of the West (on the culture of late Stalinism,
see Jeffrey Brooks, *Thank You, Comrade Stalin! Soviet Public Culture from
Revolution to Cold War* [Princeton: Princeton University Press, 2000], 195-232),
Vavilov's contention that "the conservation of mass [or matter] in chemical
transformations, which in the nineteenth century became the fundamental
law in chemistry," had been mistakenly ascribed to "the French chemist
A. Lavoisier," whereas in truth, Vavilov avers, "Lavoisier himself never laid
claim to discovering this law. The honor of its discovery belongs to M. V. Lomonosov," may not seem an extraordinary pronouncement. Vavilov's
position as the president of the Academy of Sciences, however, as well as
his being a respected historian of science, gave his words extra credence (or
perhaps better put, it imbued them with authority). Referring the reader
to the same fragment from Lomonosov's letter to Euler that Menshutkin
highlighted, as well as his 1760 speech, Vavilov also credits Lomonosov
with, among other things, outlining the principle of the "conservation
of energy." Alexei Kojevnikov reasons that in his *Pravda* article Vavilov
"masked behind a baroque phraseology the absence of a definite claim for
Lomonosov's priority in discovering the laws of matter and conservation
of energy." Kojevnikov, *Stalin's Great Science*, 181. In short, Vavilov, caught
up in the "rise of the nationalist tide" in postwar Soviet society, fulfilled his
obligation by making overstated claims like these on behalf of Lomonosov.
This last point is echoed in Tropp, "Fizika i khimiia," 34. In later, politically
safer, times, the topic of Lomonosov and Lavoisier could be treated with
more judiciousness. See, for example, Petr Kapitsa, "Lomonosov and World
Science," in *Collected Papers of P. L. Kapitza*, ed. D. Ter Harr (Oxford: Pergamon
Press, 1967), 168-84. Kapitsa suggests that Lomonosov's "discovering of
the law of the conservation of matter is now well studied and it has been
fully established that Lomonosov was the first to discover it. Lomonosov's
experiment was analogous to the celebrated experiment of Lavoisier, though
Lavoisier's was 17 years later. I shall not repeat in detail all this history, most
people know it" (ibid., 177). Thus acknowledging Lomonosov's apparent
discovery, although eliding any exegesis of it, Kaptisa could, to an extent,
move forward to other subjects. It can be added that Kaptisa's essay is a fair
reading of Lomonosov's legacy; complimentary, but not devoid of criticism,
especially as concerns Lomonosov's lack of rigorous mathematical training
(ibid., 180).

to contestation; nevertheless, it seems pointless to pursue it even further.

How does Menshutkin, finally, pausing to look back not only on Lomonosov but also on his own forty years of painstaking study, place Lomonosov both in the history of science and in the history of Russia?

> Nowadays more than anything else we esteem Lomonosov as an outstanding philosopher and thinker. Already as a student he guessed the fundamental theme of research that would most promote the development of physics and chemistry: the study of the tiniest particles, from which all bodies are composed, and of the attributes of those particles. Reducing all occurrences [in nature] to the attributes of the particles out of which bodies are made up, he himself came to some very remarkable deductions and predicted the general conditions and the trajectories of the development of physics and chemistry to our own time. In many other sciences he also suggested very interesting thoughts not vindicated until many years afterward. His many-sided genius is manifested everywhere, and everywhere he was ahead of his time by years, by decades, and even by centuries...[74]

Such is Menshutkin's summing up of Lomonosov, and it differs not at all from the conclusions he offered in his 1911 speech, nor do those conclusions diverge sharply from how Lomonosov was represented in the earlier, pre-Menshutkin, investigations that he rather peremptorily dismissed.

Menshutkin's lasting addition to the Lomonosov legend concerns, ironically, his attention to the most deleterious consequence of Lomonosov's isolated labors in the eighteenth century. At a time when the state of chemistry and physics was in Menshutkin's reading so benighted (particularly within Russia), and where the level of work at the Academy had fallen off since the days of Euler and Bernoulii, Lomonosov's name inevitably became detached from his achievements. Both contemporary and later generations of scientists, who "not comprehending the meaning [or portent] of his

---

[74]   Menshutkin, *Zhizneopisanie Mikhaila Lomonosova*, 230-31.

work in chemistry and physics ... believed it was not deserving of special attention," ignored it, much to Russia's misfortune. But then how could they be mindful of Lomonosov's deeds, for even "today" we are only just becoming aware, thanks mainly to the author's own efforts, of what Lomonosov attained as a chemist and a physicist. We have here a wonderful conjunction of elevating Lomonosov's status, while also promoting Menshutkin's and that of chemistry and physics.

There was one figure, however, who could grasp Lomonosov's true scientific worth, "who entirely appreciated him, who understood all the importance of what he accomplished and who was privy to all the details of his scientific thought," and that was Leonhard Euler.[75] As a universally respected mathematician and natural philosopher, and someone who was also so closely identified with Russian science, Euler remained the scientist enlisted most frequently to shape Lomonosov's image.[76] While Menshutkin quotes at some length from Staehlin's memoir concerning Lomonosov's forceful character, and Pushkin remained a powerful cultural resource from whose writings Menshutkin also plumbed positive references,[77] Euler appeared to be a singularly perceptive judge of Lomonosov's skills, as well as a sure influence over him. Other than a mention

[75]   Ibid., 231.

[76]   Euler's reputation never perceptibly dimmed in the Soviet era; quite the contrary, even at times of often aggressive efforts to cleanse the history of Russian science from outside influences, he remained the "founding father" of mathematics in Russia. Underscoring his posthumous prestige, the two hundred and fiftieth anniversary of his birth was celebrated in 1957 with elaborate commemorations in Leningrad. That same year Euler's body was moved from the Lutheran section of the Smolensk cemetery to the graveyard adjoining the Alexander Nevskii Monastery, and re-interned in an honored place "close to the gravestone of Lomonosov." A. N. Petrov, "Pamiatnye eilerovskie mesta v Leningrade," in Lavren'tev, Iushkevich, and Grigor'ian, *Leonard Eiler: sbornik statei*, 603.

[77]   On Staehlin's and Pushkin's characterizations of Lomonosov, see Menshutkin, *Zhizneopisanie Mikhaila Lomonosova*, 226-35, passim. 1937 was the year of the vast Pushkin Jubilee, so associations between him and Lomonosov were ubiquitous in the literature.

of Euler's few brief comments on Lomonosov's scientific talents, however, the exact nature of that influence is not explicated.

Even when expressing the most fervent approbation of Lomonosov's standing in Russian history, there is an air of melancholy surrounding Menshutkin's narrative, for despite Lomonosov's exertions there were insurmountable odds that he was unable to overcome. That he was seemingly so far ahead of his time was an obstacle that he could hardly hope to triumph over. Irrespective of his temporary eclipse in the annals of world science, the fact remains that if "Lomonosov were to find himself among us, he would discover thousands of researchers developing the theme which he always put forward as essential to the conception of matter: the study of the 'the insensible particles, which constitute bodies,' by the use of the methods of physics, mathematics, and chemistry."[78]

More prosaic difficulties, such as Lomonosov's bids to reorganize the Academy of Sciences, also proved to be insoluble challenges to him. That this venture was likely motivated as much by worries over his own eroding status within its administration during the last years of his life as by care for its maintenance is of little import, for his zeal was spent advancing learning in Russia. Echoing Lomonosov's lament to Staehlin, which Menshutkin reprints,[79] was another dirge by Lomonosov: "'I suffer because I am trying to defend the work of Peter the Great, in order that the Russians may learn, and in order that they may then reveal their quality [or dignity] ... I do not grieve about death: I have lived and I have suffered, and I know that for me the children of the fatherland will mourn.'"[80]

---

[78]   Ibid., 235-36.

[79]   Ibid., 235.

[80]   Ibid., 223. This is from an unfinished report (apparently composed by Lomonosov between 26 February and 4 March 1765) that was to have been dispatched to Catherine II, which describes the "circumstances that hinder the work of Lomonosov in the Academy of Sciences." Lomonosov, *PSS*, vol. 10, 357, 764-66. This is one of many documents or letters in the same rather choleric and self-pitying vein that he authored over the course of his tenure at the Academy.

Our author esteems Lomonosov's pride in his own attainments. Honor, acceptance by authoritative figures, and the search for status were important motifs in all the biographical constructions of Lomonosov's life, largely, it would appear, because it so deeply reflected the memoirist's own contemporary concerns.[81] Noted in the introduction to this work is that Menshutkin ended his biography by including a passage from Lomonosov's translation of Horace's *Exegi monumentum*. Menshutkin's inclusion of this is an apt metaphor for not only Lomonosov's own legacy, but also his own role in memorializing Russia's first scientist.

The year 1937 was another productive one for Menshutkin, for in addition to the biography, and his continuing, and copious, translating and editing work, he assisted L. B. Modzalevskii in compiling a thorough guide to Lomonosov's manuscript holdings at the Archive of the Academy of Sciences in Leningrad.[82] It would prove to be vital in the composition of what will likely remain the definitive version of Lomonosov's collected works (issued by the Academy of Sciences in eleven volumes, 1950-83).[83] Despite a voluminous array of writings published about Lomonosov since Menshutkin's death,[84] his work as the interpreter of Lomonosov's

---

[81] For an interesting comparison, observe the renaissance of interest in the life of Vladimir Vernadskii during the 1960s and 1970s in the Soviet Union, a development that Kendall Bailes incisively describes as an effort by Soviet intellectuals "to strengthen in Russian culture an ideology of professionalism, one with strong native Russian roots, which will help to protect their freedom of inquiry, i.e., their freedom to debate and disseminate ideas without arbitrary interference by political authorities." Bailes, *Science and Russian Culture in an Age of Revolutions: V. I. Vernadsky and His Scientific School, 1863-1945* (Bloomington, IN: Indiana University Press, 1990), IX.

[82] Modzalesvskii, *Rukopisi Lomonosova v Akademii nauk.*

[83] An examination of the volumes dealing with physics and chemistry (numbers 1-4) demonstrates the editors' dependence on Menshutkin for both commentary and translations.

[84] Among them a slightly expanded version of his 1937 biography (see B. N. Menshuktin, *Zhizneopisanie Mikhaila Vasil'evicha Lomonosova*, 3rd ed., P. N. Berkov, S. I. Vavilov, and L. B. Modzalevskii, eds. [Moscow-Leningrad, 1947]). The additions in this work were mainly those of Berkov,

lasting achievements in the sciences remains conspicuously relied on. Such subsequent fixtures in the study of Lomonosov as Vavilov, Morozov, Pavlova, Raskin, Solov'ev, and Karpeev have all depended to a great extent on Menshutkin's vision of Lomonosov's science.

It is now a century since Menshutkin encountered the imagery of Lomonosov as the father of Russian science and learning, and he left that imagery greatly reinforced. His essential contribution to it was in providing a more precisely elaborated scholarly apparatus elucidating (and creatively updating) Lomonosov's corpuscular philosophy. This was initially aimed at offering chemists and physicists (or physical chemists if you prefer) an established symbol that might prove useful in their own long efforts to securely validate their status in Russian society. This was only a part of the process of a maturing professional intelligentsia, but it was a vital part.

The enormous respect accorded to the "idea" of science in the Soviet Union allowed for the continued exploitation of Lomonosov's representation as a "founding father" of science. Furthermore, the "peasant" and "Russian patriot" contours of Lomonosov's biography easily meshed with both the quasi-Marxist criteria that the lives of iconic figures had to be made to conform to and the more assertive Russian nationalism that became ever more pronounced in the late 1940s, to make of him, for the cultural authorities, an unsurpassed symbol of Russian (and Soviet) progress in the face of apparently persistent external disparagement and threats.[85] Efforts

---

and reflect his interests in eighteenth-century Russian literary life. As for the wider bibliography on Lomonosov, the numbers go into the several thousands; with considerably more than half issued since the late 1930s (see footnote 5 in the Introduction above for more on this). The 1961 Lomonosov Jubilee, in particular, saw a spectacular surge in the literature.

[85]   Intense campaigns were conducted, particularly in the early years of the Cold War, to exalt the Russian scientific past. One of the more visible by-products of this process was the attempted creation of a Soviet pantheon of science heroes, a process compellingly described by Krementsov: "Countless biographies of founding fathers published in the late 1940s and early 1950s resembled the *Lives of the Saints*. All were constructed in accordance with the same plan: the founding father of every field, as it happened, had been (with very few exceptions) a Russian; he had been a materialist; he had sympathized with socialism, worked fruitfully for the

Soviet stamp commemorating
Lomonosov, 1956

to recast Lomonosov in the rather nebulous Soviet mold demanded by the new ideological dictates profoundly compromised the sustainability of his historical reputation. The myth of Lomonosov eventually became so corrupted by extravagant display, centralized manipulation, and, perhaps most importantly, by its intimate association with the failed Soviet "experiment" itself that it may have irretrievably lost the irrefutable worth that generations of Russians attached to it.

---

common good, and criticized foreign science (and had often been defamed, abused, mistreated, or insufficiently appreciated by it). If the father had died before the revolution, he had struggled against (or at least been unsympathetic to) the tsarist government." Krementsov, *Stalinist Science*, 223. It is manifestly easy to see how Lomonosov's biography dovetailed perfectly with such prescriptions, even with the "socialism," which could be finessed. Besides Krementsov, the following studies are markedly informative on Soviet science and ideology in the immediate post-war period: Ethan Pollock, *Stalin and the Soviet Science Wars* (Princeton: Princeton University Press, 2006); Paul Josephson, "Stalinism and Science: Physics and Philosophical Disputes in the USSR, 1930-1955," in *Academia in Upheaval: Origins, Transfers, and Transformations of the Communist Academic Regime in Russia and East Central Europe*, ed. Michael David Fox and György Péteri (Westport, CT, 2000), 105-38; Gennadii Gorelik, *Sovetskaia zhzin' L'va Landau* (Moscow, 2008); David Holloway, *Stalin and the Bomb: The Soviet Union and Atomic Energy, 1939-1956* (New Haven: Yale University Press, 1994); Kojevnikov, *Stalin's Great Science*; and A. S. Sonin, *Fizicheskii idealizm: istoriia odnoi ideologicheskoi kampanii* (Moscow, 1994), 87-204. Finally, V. D. Esakov gained access to still-partially-restricted archives and compiled *Akademiia nauk v resheniiakh Politbiuro TsK RKP(b)-VKP(b): 1922-1952* (Moscow, 2000). Esakov's volume illustrates the importance that Soviet authorities at the highest level placed on both science administration, and on the history of science, from the early, building, years of socialism to "high Stalinism."

AFTERLIFE OF THE MYTH

*D*isquieted by the persistence of the "myths of anticipation and other forms of mythical history" that often define the genre of scientific biography, accounts that "typically present the hero as a genius struggling against a stupid contemporary world that placed every kind of obstacle in the way of his brilliant ideas; ideas that are brilliant because they anticipated, or can be read into, modern knowledge," Helge Kragh asserts that "it is obviously the duty of the historian to puncture myths, where these can be located."[1] Eliding the rather anachronistic positivism of such an approach to myth,[2] it can also be argued that the overturning of such symbols,[3] however cleverly accomplished, would erode even

---

[1]   Helge Kragh, *An Introduction to the Historiography of Science* (Cambridge: Cambridge University Press, 1987), 168-69. As additionally noted by Kragh, in a point that describes much of the literature on Lomonosov: "Such obstacles will often not have any authentic basis in fact but will merely be a means of strengthening our admiration for the hero (if he overcomes them) or of excusing his lack of success (if in spite of everything, he does not overcome them)." See also Kragh, "Received Wisdom in Biography: Tycho Biographies from Gassendi to Christianson," in Söderqvist, *History and Poetics*, 121-33.

[2]   Less archaic arguments are propounded in Nicolaas A. Rupke's *Alexander von Humboldt: A Metabiography* (Chicago: University of Chicago Press, 2008

[3]   Such efforts seem not to have shaken enthusiasm for myriad worthies of seemingly enduring stature. The best example would be Newton, for it would be difficult to identify another scientist that has been subject, at least over the past few decades, to more concerted scholarly attention, and often of a very critical kind. But in contradiction to Rupert Hall's contention that as the result of both a more sophisticated approach toward biographical writing (here he specifies Frank Manuel's *A Portrait of Isaac Newton*) that has attempted to get at Newton's character and the opening of relevant archives, "myth and prejudice have been dispelled ... and the old view of

further that interest in the history of science that devotees of this field of study have always tried to promote. It seems more relevant to try, as this inquiry has attempted, to understand why a particular myth was fashioned in the first place.

Shorn of its later Soviet ideological embellishments, the mythology of Lomonosov essentially remained what it had been at Menshutkin's passing, an elaborate imagery celebrating the father of Russian science. Representations of Lomonosov by Staehlin, Verevkin, and Novikov, structured around legendary accomplishments and tireless struggles, continued to be fundamental to the construction of his idealized biography. Later evaluations by Murav'ev and Radishchev (the latter's assessment utilized quite selectively) along with Severgin's, Pushkin's, and Pogodin's fulsome praise, further enshrined Lomonosov's status as an intrepid fighter for the dissemination of enlightenment in Russia. His roles at the Academy of Sciences and in founding Moscow University were the central tropes in this element. Lastly, Menshutkin offered, however tendentious it may have been, a scholarly investigation of Lomonosov's science itself, while never losing sight of the necessity of subsuming the minutiae of Lomonosov's chemical and physical treatises into the established narrative of Lomonosov's life. Largely due to Menshutkin's labors, Lomonosov's name was appropriated by a host of scientific disciplines, especially chemistry and physics

---

him as superman and national hero has vanished" (Hall, *Isaac Newton*, 192), to this observer, the mythology around Newton appears as robust as ever. Hall's statement also raises the more important point of why such efforts at making Newton more human should have been an objective. Hall himself belies his own inferences that Newton hagiography is a thing of the past by remarking that "the reward of scholarship, an unvarnished Newton, besides gratifying the desire to vindicate truth, enables his magnificent, outstanding intellect to be better understood than ever before." Patricia Fara, *Newton: The Making of Genius* (London: Macmillan, 2002), through an examination of print and visual representations, traces the extraordinary ascent of Newton's reputation over the past three centuries. A beautifully illustrated account of Newton's emergence as the personification of modernity is Mordechai Feingold, *The Newtonian Moment: Isaac Newton and the Making of Modern Culture* (Oxford: Oxford University Press, 2004).

but also geology, mineralogy, geography, and so forth, which began to emerge with force in Russia at the end of the nineteenth century.[4]

Whether the intent was to elevate the status of the natural philosopher in the eighteenth and nineteenth centuries or to impress upon the regime (or on society, however defined) the continued relevance of the Academy of Sciences and Moscow University, or at the end, to further the socioprofessional position of the chemist and physicist in late Imperial Russia and in the early years of the Soviet Union—in short, whatever the motivations of the individual mythmaker—the overall effect was to insist on the importance of science, and for many the more practical need to support science, in a modernizing nation. Notwithstanding the problematic definitions of nation or nationalism, or for that matter of modernization, from its earliest manifestations the Russian character of Lomonosov's biography was always placed in opposition to some sorts of foes, initially not necessarily foreign, who had attempted in some manner to undermine his efforts on behalf of his people. Substitute Soviet for Russian, and a far more heated emphasis on perceived antagonists, nearly all of whom were non-Russian, along with a commensurate focus on the particularities of a supposedly unique Russian scientific tradition, and we have many of the additions to the Lomonosov myth over the past half-century.

---

[4]   Pnina Abir-Am describes a certain process structuring the development of scientific commemorations, starting with those enshrining "great minds and institutions," which became common at the beginning of the previous century, moving on to the celebration of discoveries that has become commonplace in scientific life over the past half-century, and ending in the present-day "commemorative mania" that is often the public face of science. This could, however, easily result in a narrowing of the scope of the "hero-scientist's" symbolic reach, for as she compellingly points out, "Ironically, while personalized grandeur, however subjective, enables a wider identification with more social groups, more objective commemorative objects (such as disciplines and institutions) seem to appeal primarily to scientific elites concerned with controlling those objects' public image as contributors to scientific progress and social welfare" (Abir-Am, in Abir-Am and Elliot, *Commemorative Practices in Science*, 29). Conceptually, if not strictly chronologically, Abir-Am's schematization parallels the past framing of organized celebrations in the Lomonosov cult.

This is not to say that Lomonosov was not richly shrouded in the language of Soviet rhetoric, for indeed his biography became a starting point for extensive efforts to pinpoint the initial appearance in Russian culture of materialist-oriented, deeply anti-clerical, enlightened thinkers whose relationship with the state was fraught with political conflict. So Lomonosov as the diffuser of enlightenment in Russia—an indispensable component of the myth over the past two centuries—was repositioned as a symbol of enlightenment for the Soviet era, someone whose attributes might be emulated or internalized. The dubious nature of such historical terminology need not be emphasized, and its farcical quality in post-Soviet Russia is even more self-evident.

Previous chapters have demonstrated the valuable work Soviet scholars have produced on Lomonosov. The best of these studies are, however, marked by the narrowest of monographic styles, which in politically more sensitive times provided a relatively safe scholarly route for historians of Russian science and culture. These studies provide invaluable details, particularly bibliographic, for the interested student, although their effects on the reception of Lomonosov's image seem to have been highly muted, for it was the unblemished biography (in the vein of sanctification), not the details of their treatises, that continued to capture the attention and imagination of later generations.

What Soviet-era hagiographers attempted was to subsume representations of Lomonosov into an all-embracing cultural mission aimed at creating a "New Soviet Man," one imbued with a revolutionary communist consciousness and an accompanying passion to surmount all challenges or rather "unmask" all enemies. Governed by the demands of *Partiinost'* ("party mindedness" or "partyness"),[5] which entailed the presumptive subordination of scientific and cultural life to party dictates, elusive though they may have been at times, an ever more univalent depiction of Lomonosov's life evolved.

---

[5] On the philosophical foundations of *partiinost'* and its relationship to science, consult Joravsky, *Soviet Marxism and Natural Science*; and idem, "Soviet Views on the History of Science."

Although Lomonosov's biography had been established well before the onset of Soviet cultural experimentation, the contours of his life were seen as a convincing model for what was hoped would be indefatigable, practical-minded, patriotic (for a patriotism also equivocally defined), and cultured new Soviet generations.[6] Of course this particular vision, along with the wider societal eschatology, never progressed beyond intent. When the Soviet project failed, as it did in terms of the domestic support the ideology could marshal by the early 1970s, then what remained in addition to the truly indigestible amount of writings given over to Lomonosov was the residue of a more public style of veneration; one that distinguished Soviet displays toward officially approved cult figures. This praise, which assumed a staggering scale, was orchestrated quite without exception by the state.

Commencing around 1940, with the rechristening of Moscow University in Lomonosov's honor, concerted attempts to glorify Lomonosov were launched in numerous communal spaces. This included not merely ubiquitous statuary across the Soviet Union but the requisite naming of towns and villages (most famously Oranienbaum, west of Leningrad, in 1948), along with metro stations, schools, streets, and in a none too subtle effort to link Lomonosov with Soviet scientific achievements (and pretensions), a crater on the moon, after him. Innumerable like examples in several genres could be cited.[7]

Two disparate tributes to Lomonosov are essential to single out. The first, the Lomonosov Museum (an initiative of Sergei Vavilov) in St. Petersburg, established in 1949 atop the *Kuntskamera* (Russia's oldest museum, founded by Peter the Great as a library and chamber of curiosities in 1714), was meant to serve as a temple for the study of Lomonosov, and for a time did perhaps fill that

---

[6]   A case in point is B. G. Kuznetsov, *Tvorcheskii put' Lomonosova* (Moscow, 1961). Kuznetsov's widely-cited work was published to coincide with the 250th anniversary of Lomonosov's birth.

[7]   Chenakal, *Lomonosov v portretakh, illiustratsiakh, dokumentakh*, 229-94, offers an introduction to the topic.

function.[8] Correspondingly notable, and arguably the apogee of popularized attempts to canonize Lomonosov as a precursor to the New Soviet Man, was the 1955 film *Mikhailo Lomonosov* (directed by Mikhail Shapiro).[9] The film is a splendid illustration of how *partiinost'* operated in practice.

Throughout *Mikhailo Lomonosov* there is a predominant, relentlessly chauvinistic image of Lomonosov's crusade to advance Russian science despite the barriers raised by treacherous foreigners, self-serving gentry, and ignorant clerics. Euler and Richmann, despite not being Russian, are accorded respectful treatment, though they are shown as functioning purely in the radiance of Lomonosov's achievements. Müller, Schumacher, and nearly all other "foreign" academicians are cast as utterly villainous. Several similar cinematic treatments of scientists were produced from the late 1940s to the mid-1950s, and without exception they offer formulaic accounts of the struggles and victories of Russian scientists against ignorance, avarice, and past sycophancy before "non-native" science (classics of the "genre" are *Michurin*, 1948; and *Academician Ivan Pavlov*, 1949).While these films might induce derisive laughter today due to their crude ideological biases, they offer compelling viewing.[10]

8   The Lomonosov Museum, which organizationally exists under the auspices of the Russian Academy of Sciences, is headed by Tatiana Moiseeva. It remains a center for publishing items relating to Lomonosov, however remotely. For histories of the museum as well as descriptions of its holdings, see V. L. Chenakal, ed. *Muzei M. V. Lomonosova v Leningrade* (Leningrad, 1967); and I. V. Breneva and T. M. Moiseeva, *Muzei M. V. Lomonosova: putevoditel'* (St. Petersburg, 1995): and M. F. Khartanovich and N. P. Kopaneva, eds. *Mikhail Vasil'evich Lomonosov. K 300-letiiu so dnia rozhdeniia: po materialam Muzeia M. V. Lomonosova* (St. Petersburg, 2011).

9   *Mikhailo Lomonosov*, dir. M. Shapiro (Leningrad, 1955), videocassette.

10   Far less curious is a nine-part movie biopic *Mikhailo Lomonosov*, (dir. Aleksandr Proshkin, Moscow, 1984-86) that was shown on Soviet television in the mid-1980s. With its outsized cast, familiar leading actors, and large budget, Proshkin's film is superficially more sophisticated than the Shapiro production. Even so, it reenacts the same clichés, except more statically and at several times the length of the earlier film. Both the 1955 and 1984-86 versions were shown in November 2011 (the tercentenary of Lomonosov's birth) on Russia's ORT Channel One. The films were also presented at

Both the Lomonosov museum and *Mikhailo Lomonosov* resolutely, without a suggestion of nuance, join Lomonosov to Peter the Great's "modernizing" efforts—although, given past discourses in that direction, the association was long since established. Such clumsy efforts at deification were the most visible feature of Soviet reverence towards Lomonosov, and they were infused with potentially troublesome implications for the myth it was meant to celebrate. For functioning in conjunction with the literary overexposure, they contributed in great measure to the cynicism that saturated such prescribed esteem during the waning years of the Soviet Union.[11] Whatever the original impulse for such massive

"Lomonosov Day" festivities organized by the Russian Centre at Yerevan State University. About a dozen students from the Faculty of Russian Philology discussed the films. For anyone but enthusiasts—none of whom, besides me, was in attendance at the aforementioned screenings—Proshkin's film is merely tedious. Lacking even the comical heavy-handedness of Shapiro's *Lomonosov*, it reflects an ideologically spent late Soviet society. On the socialist project and the crafting of historical films during the Stalin era and beyond, see Evgeny Dobrenko's perceptive *Stalinist Cinema and the Production of History: Museum of the Revolution* (New Haven: Yale University Press, 2008).

[11] Pierre Nora's distinction between "imposed symbols" and "constructed symbols" in French culture is a useful organizing device in the study of myth. Imposed symbols, such as the pantheon and the Eiffel tower, are where "symbolic and memorial intention is inscribed in the object itself," and which are often quickly transformed into "official state symbols." Constructed symbols, for example Joan of Arc and Descartes, emerge as the result of "unforeseen mechanisms, combinations of circumstances, the passage of time, human effort, and history itself," which turn them "into important and durable symbols of Frenchness." Nora, preface to *Realms of Memory: The Construction of the French Past*, vol. 3, trans. Arthur Goldhammer (New York: Columbia University Press, 1998), X. When myths become utterly annexed by the state, or by quasi-state institutions, they of course cease to grow, and their symbolic worth is gradually undermined. This would be the case with Lomonosov, and was very nearly the case with Pushkin. See Levitt, *Russian Literary Politics*, 147-75; Sandler, *Commemorating Pushkin*; and Debreczeny, *Social Functions of Literature*, 223-46, for both late Tsarist and Soviet efforts to control Pushkin's legacy. For a closer study of Pushkin in the Soviet era, see Molok, *Pushkin v 1937 godu*. For concerted Soviet efforts—ultimately failed efforts—to generate an effective myth around the figure of Pavlik Morozov, see Catriona Kelly, *Comrade Pavlik: The Rise and Fall of a Soviet Boy Hero* (London: Granta, 2005). Party/state-fostered

eulogizing efforts, its effects on Lomonosov's historical eminence
were rather less than were intended. It is not paradoxical that the
strength and meaning of the Lomonosov myth began to erode at
the very time when it received its most fulsome praise, for it was
also monochromatic homage, of the type that almost begged, when
noticed at all, for a barbed riposte.

In a richly revealing anecdote, not atypical of those which
were widely circulated during the later decades of the Soviet Union,
Sergei Dovlatov writes of the completion by a team of molders
(or "stone carvers") of a marble relief of Lomonosov intended for
placement in a new metro station in Leningrad:

> Lomonosov was depicted in some kind of suspicious-looking
> robe. In his right hand he held a paper scroll; in his left,
> a globe. The paper, as I understand it, symbolized the creative
> spirit, and the globe—science.
>
> Lomonosov himself looked well-fed, effeminate, and
> slovenly. He resembled a pig. During the Stalin years that's
> how they portrayed capitalists. Evidently Chudnovskii [the
> sculptor] wanted to confirm the primacy of matter over the
> spirit.[12]

---

adoration of Stalin is outlined in Jan Plamper, *The Stalin Cult: A Study in the
Alchemy of Power* (New Haven: Yale University Press, 2012). The impressive
Peter the Great imagery in Russian life has been touched on; striking also is
the evolution of Lenin iconography in the Soviet Union. The quite religious
veneration accorded the Bolshevik leader is investigated in Nina Tumarkin,
Lenin Lives; as well in Olga Velikanova, *The Public Perception of the Cult of
Lenin Based on Archival Materials* (text in Russian) (Lewiston, NY: The Edwin
Mellen Press, 2001). Worth stressing is that the origins of the Lenin "cult"
evidences both constructed and imposed elements, a cross-fertilization that
also applies to Pushkin, which Nora's mechanism does not easily allow for.
It is perhaps too early to gauge with precision what post-Soviet views will
be on these figures; representations of Pushkin and Peter the Great were
never as utterly subsumed to the grotesqueries of later mythmakers as
were those of Lomonosov and Lenin; considerable "independent" content
remained.

12   Sergei Dovlatov, *Chemodan* (Tenafly, NJ: Ermitazh, 1986), 23-24. Dovlatov,
who emigrated from the Soviet Union to New York in 1978, was first
brought to literary eminence by Joseph Brodsky's praise. For more than
three decades his writings have enjoyed a posthumous surge of popularity
in Russia and elsewhere in the former Soviet Union.

It is not the contempt in Dovlatov's description that strikes
the reader, but rather the indifference. Completing the piece,
Lomonosov's visage took on its familiar guise,[13] and it was not
a pleasant one: his "image was becoming clearer. And, it must be
remarked, even more repellent."[14] Finally, as they finished the head
and placed it into position in the metro, it could only be judged
that "from afar Lomonosov looked more tolerable." Although this
is an apocryphal account, much like Dovlatov's tale of an irreverent
unveiling of a statue to Lenin in Cheliabinsk,[15] it does convey the
seeming purposelessness of the public imagery of Lomonosov. But
even as an object of such anecdotal humor, at which Soviet citizens
excelled, Lomonosov gradually faded into oblivion.[16]

Although Dovlatov's tale was merely a pretext for an attack
on the sclerotic conditions of Soviet life generally, and on the
decrepitude of Leningrad officialdom in particular, he concludes
with an amusing, yet also salient, aside: "Our Lomonosov was
removed in two months. Leningrad scientists wrote a letter to
a newspaper, complaining that our sculpture belittled [or humi-

---

[13]   Two centuries of visual reproductions have rarely strayed from the original
         eighteenth-century portrayal of Lomonosov's puffy-cheeked countenance.
         On the history of Lomonosov portraiture, see Glinka, *Lomonosov (opyt
         ikonografii)*. An impressive statue of Lomonosov was erected during the
         1986 Lomonosov Jubilee on a conspicuous site near the *Kuntskamera*.
         The convoluted efforts to select an acceptable model are covered in
         Rytikova, "Obraz M. V. Lomonosova v monumentalynikh zamyslakh
         Leningradskikh skul'ptorov." This Lomonosov is missing his customary
         wig: apparently wigs better suit a foppish servitor than an industrious
         scientist.

[14]   Dovlatov, *Chemodan*, 25-26.

[15]   Ibid. 21.

[16]   Of course, humor alone does not undermine the power of myth; indeed, it
         probably adds to its strength through the inclusion of a certain multivalence.
         Epigrams, puns, jokes, and the like, which played off Lomonosov's "low
         origins" and presumed fondness for heavy drinking, were a staple of
         eighteenth-century literary polemics, and allowed, if unintentionally,
         a more human element to enter Lomonosov's later biography. It is the lack
         of humor that, among other symptoms, probably indicates the decline of
         a myth.

liated] a great figure."[17] So the scientific community, as complicit as the cultural and political authorities in these rituals, felt it incumbent on itself to defend Lomonosov's standing. Although this was evidently not intended by Dovlatov as anything more than a further comical assault on Soviet norms, it alludes to one way later generations might remember Lomonosov.

If the mechanisms that select and subsequently shape the formation of myths are difficult to locate with precision, and indeed are best glimpsed by an extended examination of both representative tropes and responses to them, then to speculate on the myth's future is equally fraught with conceptual pitfalls. It need be recognized at once that it was neither scholarly attention (as was suggested by Helge Kragh and most historians of science) nor related efforts at deconstruction that undermined Lomonosov's symbolic meaning. Moreover, the notion that it was mainly the official nature of memorializations that bears much of the responsibility is contested by the adoration still accorded certain figures, most palpably Pushkin, in Russian culture.[18]

Something far more consequential is at fault. Unlike with Pushkin's biography, and more aptly for a study of scientific myths, biographies of individuals such as Newton, Descartes, Galileo, Copernicus, and Franklin, it might well be that with Menshutkin's interpretively far-reaching exhumation of Lomonosov the scientist, the mythology's ability to inspire the intelligentsia and create devotees simply exhausted itself. Add to this that with the late collapse of the Soviet experiment, entailing of course the undermining of the official culture that had so vigorously promoted

---

[17]   Dovlatov, *Chemodan*, 31.

[18]   Although an obvious objection might be that Pushkin represents a distinct template, I would nonetheless argue that the comparison is instructive. The bicentennial of Pushkin's birth was celebrated with great fanfare in 1999. For a survey of some of the literature that greeted the commemorations, see *Moskovskii Pushkinist* (a series launched in 1995) and nearly every issue of *Novoe literaturnoe obozrenie* (particularly its review of new books) for the several years succeeding the jubilee. On how Pushkin's image has been utilized in recent Russian nationalist discourse, interesting is Wendy Slater, "The Patriots' Pushkin," *Slavic Review* 58, no. 2 (Summer 1999): 407-27.

adoration of Lomonosov as the *syn velikogo russkogo naroda*, a largely indifferent public is no longer faced with the requirement to even superficially consider Lomonosov.[19]

Lomonosov in post-Soviet Russia has slipped from being a dazzlingly resonant symbol of Russian scientific triumphs, to being one who, although still the eponymous father of Russian science and learning, arouses no debate. Only perfunctory gatherings conducted by the remaining bearers of the torch, most of whom are to be found at either at the Institute for the History of Science and Technology or at the nearby Lomonosov Museum, seem in the offing.[20] Thus, even in academia, the last bastion of antiquarianism, there are worrying signs for the dwindling enthusiasts (or disciples) of Lomonosov. For, unlike the resurgence of often-superb scholarly interest that has greeted a wide variety of figures from the Russian past, both those neglected and those the subject of long attention who are now being reevaluated to meet changed times,[21] Lomonosov's story has suffered a precipitous fall in attention. Systematic searches of recently published and forthcoming publications, even in this past tercentenary "jubilee" year (2011),[22] indicates a striking

---

[19]   Assessing the acute "desacralizaton" of the Decembrists over the past twenty years, Ludmilla Trigos hesitatingly makes the obvious inference that the "Decembrists' association with the Soviet regime's legitimatizing myths may well have tainted them beyond re-integration": Ludmilla A. Trigos, *The Decembrist Myth in Russian Culture* (New York: Palgrave Macmillan, 2009), 182.

[20]   Lomonosov's name still receives obligatory tributes such as that given by the president of the Russian Academy of Sciences, Iurii Osipov, at the Academy's two hundred and seventy-fifth anniversary ceremonies in 1999 (see Iu. S. Osipov, *Akademiia nauk v istorii Rossiiskogo gosudarstva* [Moscow, 1999], 25-31, passim), although now the accolades are not simply dispassionate. They also no longer eclipse acknowledgement of the deeds of other, particularly "foreign," eighteenth-century academicians.

[21]   An unfortunate exception to this is Aleksandr Radishchev, whose biography was severely distorted during the Soviet period, when he was, in a vulgar treatment, cast into the mold of a "revolutionary democrat." Radishchev's writings on a wide variety of topics, especially his *Journey from Petersburg to Moscow*, deserve, nay require, extensive revisiting.

[22]   To commemorate the 300th anniversary of Lomonosov's birth, the State

dearth of innovative scholarly interest in Lomonosov's scientific biography.[23]

A perhaps not unlikely fate for Lomonosov's further reputation is that the heroic elements of his biography, which were heralded by his contemporaries and then historicized with little initial impact, first by Radishchev and then in the nineteenth century, most compellingly, if also singularly, by Liubimov, will constitute its essence. While Radishchev and Liubimov were admiring of Lomonosov's symbolic value to Russians, they demonstrably elided any notion of lasting and direct influence. Whether or not the forces that placed and kept Lomonosov in the pantheon of Russian greats, which consisted of his power to enthuse later generations, will be

---

Hermitage Museum in St. Petersburg hosted a scholarly conference and exhibit in November 2011 called "Mikhail Lomonosov and the Time of Elizabeth I." This gathering, with its too-obvious echoes of the similarly-named exhibit held soon after the 1911 Jubilee, is most noteworthy for its parochialism. No international speakers delivered papers, nor was there any attempt in the lectures given to address the most noticeable lacuna in Lomonosov studies: the failure to situate Lomonosov's legacy in the vast corpus of literature on the history of European science. This, of course, would require not only encountering non-Russian language scholarship, but also incorporating it. On the Lomonosov conference, see http://www.hermitagemuseum.org/html_En/00/hm0_4_499.html. A handsomely illustrated catalogue was issued to accompany the exhibit: N. Iu. Guseva, comp. *M. V. Lomonosov i elizavetinskoe vremia: Katalog vystavki* (St. Petersburg, 2011). As impressive as the visual presentation is, the accompanying text is uninspired.

[23] Peter Hoffmann's *Michail Vasil'evic Lomonosov*, the culmination of decades of interest by the author in the Russian eighteenth century, fittingly was issued during the Lomonosov tercentenary. Hoffmann's text, a meticulously researched, authoritatively written chronicle of Lomonosov's richly varied life, offers, however, very little that is interpretively novel, especially as regards Lomonosov's scientific legacy. Neither does Hoffmann contextualize Lomonosov within the wider currents of "Enlightenment" Europe. Outside of the predicable reprints of Lomonosov's writings alongside those of some canonical texts (such as Menshutkin's 1911 biography), the most consequential Russian language publications issued for the jubilee are reference works. Specialized studies on Lomonosov's library, documents associated with his work at the Academy of Sciences, and guides to his correspondence (all cited herein) were published over the past two years. None of them can be construed as breaking new ground in the scholarship.

resurrected, is also uncertain. But with Russia undergoing many of the same trials of trying to catch up, or at least perceiving the need to draw nearer, with the "West" which have marked its development since Peter the Great's time, it is plausible that Lomonosov might again serve as an exemplar.

Nations require myths, and those of heroes, or founding figures, are commonplace.[24] Myths can tell the observer much of the history of the country and the people formulating such symbols, a need that biographies and related narrative histories would be hard pressed to match. So the question to return to is not the truth value of such representations, but rather to what use they are put. Two centuries of Lomonosov mythology have, I trust, made known much of how Russian scientists, poets, historians, and journalists, among others, have assessed the importance of science and learning to their nation, indeed how they saw the furtherance of *nauka* as vital to Russian development and national pride. With Lomonosov we are not dealing purely with the "falsification" of history (to use a favorite Soviet term in depicting so-called bourgeois historiography), although there is much of that in the iconography of his life. More significantly, it must be argued, he is a figure who long embodied ideals that many Russian thinkers sought to disseminate in their country.

[24]   On the many reiterations of the myths—principally religious, national, and state—that constitute Aleksandr Nevskii's biography, and their centuries of deployment in "Russian" discourse, see Benjamin Frithjof Schenck, *Aleksandr Nevskij: Heiliger—Fürst—Nationalheld. Eine Erinnerungsfigur im russischen kulturellen Gedächtnis (1263-2000)* (Köln: Böhlau, 2004). The Petrine and post-Petrine appropriation of Aleksandr Nevskii to imperial/state demands (commencing with chapter 5) is especially germane to the invention of comparable Russian "national heroes."

BIBLIOGRAPHY

Reference Guides

Berkov, P. N., et al., eds. *Istoriia russkoi literatury XVIII veka: bibliograficheskii ukazatel'*. Leningrad, 1968.

Beliaeva, I. M., ed. *Biblioteka M. V. Lomonosova: nauchnoe opisanie rukopisei i pechatnykh knig*. Moscow, 2010.

Chenakal, V. L., A. V. Topchiev, and N. A. Figurovskii, eds. *Letopis' zhizni i tvorchestva M. V. Lomonosova*. Moscow-Leningrad, 1961.

Fomin, A. G., et al., eds. *Materialy po bibliografii o Lomonosove na russkom, nemetskom, frantsuzskom, ital'ianskom i shvedskom iazykakh*. Petrograd, 1915.

Karpeev, E. P., ed. *Lomonosov: kratkii entsiklopedicheskii slovar'*. St. Petersburg, 1999.

Khartanovich, M. F., et al., eds. *Letopis' Rossiiskoi Akademii nauk*. Vol. 2, 1803-1860. St. Petersburg, 2002.

Klado, T. N., et al., eds. *Leonard Eiler: perepiska, annotirovannyi ukazatel'*. Leningrad, 1967.

Korovin, G. M. *Biblioteka Lomonosova: materialy dlia kharakteristiki literatury, izpol'zovannoi Lomonosovym v ego trudakh, i katalog ego lichnoi biblioteki*. Moscow–Leningrad, 1961.

------, ed. *Mikhail Vasil'evich Lomonosov: ukazatel' osnovnoi nauchnoi literatury*. Moscow-Leningrad, 1950.

Kuliabko, E. S. and E. B. Beshenkovskii. *Sud'ba biblioteki i arkhiva M. V. Lomonosova*. Leningrad, 1975.

Kuntscvich, G. Z., ed. *Bibliografiia izdanii sochinenii M. V. Lomonosova na russkom iazyke*. Petrograd, 1918.

"Lomonosovskiia torzhestva. (Bibliograficheskaia zametka.)" In *Pamiati M. V. Lomonosova. Sbornik statei k dvukhsotletiiu so dnia rozhdeniia Lomonosova*. St. Petersburg, 1911, 88-105.

Martynov, G. G., ed. *Mikhail Vasil'evich Lomonosov: perepiska 1737-1765*. Moscow, 2010.

Mezhov, V. I. *Iubelei Lomonosova, Karamzina i Krylova: bibliograficheskii ukazatel' knig i statei, vyshedshikh po povodu iubileev.* St. Petersburg, 1872.

Modzalevskii, L. B., ed. With a preface by B. N. Menshutkin. *Rukopisi Lomonosova v Akademii nauk SSSR: nauchnoe opisanie.* Moscow-Leningrad, 1937.

------, I. V. Tunkina, ed. *M. V. Lomonosov i ego literaturnye otnosheniia v Akademii nauk: Iz istorii russkoi literatury i prosveshcheniia serediny XVIII v.* (St. Petersburg, 2011).

Nevskaia, N. I., et al., eds. *Letopis' Rossiiskoi Akademii nauk.* Vol. 1, 1724-1802. St. Petersburg, 2000.

Ponomarev, S. I., ed. *Materialy dlia bibliografii literatury o Lomonosove.* St. Pe-tersburg, 1872.

Putevoditel' po vystavke "Lomonosov i elizavetinskoe vremia". St. Petersburg, 1912.

Stolpianskii, P. N. "Nekotorye dannye k biografii Lomonosova, izvle-chennye iz 'S. Peterburgskikh vedomostei' za XVIII vek." In *Lomonosovskii sbornik.* St. Petersburg, 1911, 267-82.

Svodnyi katalog russkoi knigi grazhdanskoi pechati XVIII veka, 1725-1800. 6 volumes. Moscow, 1962-75.

"Ukazatel' iubileinoi literatury o Lomonosove." In *Pamiati M. V. Lomonosova. Sbornik statei k dvukhsotletiiu so dnia rozhdeniia Lomonosova.* St. Petersburg, 1911, 106-22.

Archival Sources

Considering Lomonosov's venerated status in Russian and Soviet history, the fact that his writings have been the subject of profuse attention, and published and republished time and time again, should come as no surprise. Indeed, in three years of research in St. Petersburg, where nearly all of Lomonosov's writings are located, I discovered only one paper ascribed to Lomonosov, a poem of admittedly uncertain provenance, which escaped inclusion in the latest edition of his collected works. The main repository for Lomonosov's papers is the *Sanktpeterburgskii filial Arkhiva Rossiiskoi Akademii nauk, PFA RAN* (St. Petersburg Branch of the Archive of the Russian Academy of Sciences). Both Lomonosov's own writings (scientific, literary, and administrative) and many documents related to his work at the Academy of Sciences are located in *fonds* 1, 3, and 20. Citations to archival and documentary collections concerning Lomonosov herein are to either the *PFA RAN*, or to the *Otdel rukopisei* (manuscript division) of the *Rossiiskaia natsional'naia biblioteka* (Russian National Library) in St. Petersburg, and can be found in the footnotes.

Works Cited

XVIII vek, 26 volumes. Moscow-Leningrad-St. Petersburg, 1935-2011.

Abir-Am, Pnina G. "How Scientists View Their Heroes: Some Remarks on the Mechanism of Myth Construction." *Journal of the History of Biology* 15, no. 2 (Summer 1982): 281-315.

Abir-Am, Pnina, G., and Clark A. Elliot, eds. *Commemorative Practices in Science: Historical Perspectives on the Politics of Collective Memory* (published in *Osiris* 14). Chicago: University of Chicago Press Journals, 1999.

Ageeva, O. G. *Imperatorskoi dvor Rossii: 1700-1796*. Moscow, 2008.

Aksakov, K. S. *Lomonosov v istorii russkoi literatury i russkogo iazyka*. Moscow, 1846.

Alekseev, M. P. "Pushkin i nauka ego vremeni." In M. P. Alekseev, *Pushkin: sravnitel'no-istoricheskie issledovaniia*, edited by G. V. Stepanov and V. N. Baskakov. Leningrad, 1984, 22-173.

Andreev, A. Iu. *Moskovskii universitet v obshchestvennoi i kul'turnoi zhizni Rossii nachala XIX veka*. Moscow, 2000.

------. *Russkie studenty v nemetskikh universitetakh XVIII-pervoi poloviny XIX veka*. Moscow, 2005.

Andrews, James T. *Science for the Masses: The Bolshevik State, Public Science, and The Popular Imagination in Soviet Russia, 1917-1934*. College Station, TX: Texas A & M University Press, 2003.

Anisimov, E. V. *Dyba i knut: politicheskii sysk i russkoe obshchestvo v XVIII veke*. Moscow, 1999.

------. *Empress Elizabeth: Her Reign and Her Russia, 1741-1761*, edited and translated by John T. Alexander. Gulf Breeze, FL: Academic International Press, 1995.

------. "I. I. Shuvalov—deiatel' rossiiskogo prosveshcheniia." *Voprosy istorii* 7 (July 1985): 94-104.

------. "M. V. Lomonosov i I. I. Shuvalov." *Voprosy istorii estestvoznaniia i tekhniki* 1 (1987): 73-83.

Auburger, Leopald. *Russland und Europa: Die Beziehungen M. V. Lomonosovs zu Deutschland*. Heidelberg: Groos, 1985.

Baehr, Stephen L. *The Paradise Myth in Eighteenth-Century Russia: Utopian Patterns in Early Secular Russian Literature and Culture*. Stanford: Stanford University Press, 1991.

Bailes, Kendall E. *Science and Russian Culture in an Age of Revolutions: V. I. Vernadsky and His Scientific School, 1863-1945*. Bloomington, IN: Indiana University Press, 1990.

Banville, John. *Kepler: A Novel*. London: Secker & Warburg, 1981.

Barber, W. H. *Leibniz in France: From Arnauld to Voltaire—A Study in French Reactions to Leibnizianism, 1670-1760*. Oxford: Clarendon Press, 1955.

Barsukov, N. P. *Zhizn' i trudy M. P. Pogodina*, 22 volumes. St. Petersburg, 1888-1910.

Bartenev, P. I. "I. I. Shuvalov." *Russkaia beseda* 1, part 6 (1857): 1-80.

------, ed. *Arkhiv kniazia Vorontsova*, vol. 5. Moscow, 1872.

Barthes, Roland. *Mythologies*. Translated by Annette Lavers. New York: Hill and Wang, 1972.

Bartlett, Roger, and Janet M. Hartley, eds. *Russia in the Age of Enlightenment: Essays for Isabel de Madariaga*. New York: St. Martin's Press, 1990.

Beliavskii, M. T. *M. V. Lomonosov i osnovanie Moskovskogo universiteta*. Moscow, 1955.

------. "Petr Chelishev i ego 'Puteshestvie po severu Rossii'." *Vestnik Moskovskogo universiteta. Istoriko-filologicheskaia seriia*, no. 2 (1956): 19-47.

Belinskii, V. G. *Polnoe sobranie sochineniii*, vols. 2, 8-9. Moscow, 1953-55.

Bensaude-Vincent, Bernadette and Charlotte Vincent, eds. *Science and Spectacle in the European Enlightenment*. Burlington, VT: Ashgate, 2008.

Beretta, Marco. *Imaging a Career in Science: The Iconography of Antoine Laurent Lavoisier*. Canton, MA: Science History Publications, 2001.

Berkov, P. N., ed. *Literaturnoe tvorchestvo M. V. Lomonosova: issledovaniia i materialy*. Moscow-Leningrad, 1962

------. *Lomonosov i literaturnaia polemika ego vremeni*. Moscow-Leningrad, 1936.

------, ed. *Problemy russkogo prosveshcheniia v literature XVIII veka*. Moscow-Leningrad, 1961.

Bethea, David M. *Realizing Metaphors: Alexander Pushkin and the Life of the Poet*. Madison: University of Wisconsin Press, 1998.

Biagioli, Mario. *Galileo Courtier: The Practice of Science in the Culture of Absolutism*. Chicago: University of Chicago Press, 1993.

------. *Galileo's Instrument's of Credit*. Chicago: University of Chicago Press, 2006.

Biliarskii, P. S. *Materialy dlia biografii Lomonosova*. St. Petersburg, 1865.

Black, J. L. *G.-F. Müller and the Imperial Russian Academy*. Kingston and Montreal: McGill-Queen's University Press, 1986.

Blagoi, D. D. *Ot Kantemira do nashikh dnei*, vol. 1. Moscow, 1972.

Boas, Marie. "The Establishment of the Mechanical Philosophy." *Osiris* 10 (1952): 412-541.

Bobynin, V. V. "Rumovskii, Stepan Iakovlevich." *Russkii biograficheskii slovar'*, vol. 17 (St. Petersburg, 1918; reprint, NewYork, 1962), 441-50.

Bogoliubov, N. N., G. K. Mikhailov, and A. P. Iushkevich, eds. *Razvitie idei Leonarda Eilera i sovremennaia nauka: sbornik statei*. Moscow, 1988.

Boss, Valentin. *Milton and the Rise of Russian Satanism*. Toronto: University of Toronto Press, 1991.

------. *Newton and Russia: The Early Influence, 1698-1796*. Cambridge, MA: Harvard University Press, 1972.

Bradley, Robert E., and C. Edward Sandifer, eds. *Leonhard Euler: Life, Work, and Legacy*. Amsterdam: Elsevier, 2007.

Breneva, I. V., and T. M. Moiseeva. *Muzei M. V. Lomonosova: putevoditel'*. St. Petersburg, 1995.

Brewster, David. *Memoirs of the Life, Writings, and Discoveries of Sir Isaac Newton*, 2 vols. Edinburgh: Thomas Constable, 1855.

Bronnikova, V., ed. *Mikhail Vasil'evich Lomonosov: iz naslediia Lomonosova, slovo sovremennikov o Lomnoosove, 'pamiat' vechnaia', 'vysokii lik v griadyshchem pokolen'i...'*. Moscow, 2004.

Brooks, Jeffrey. *Thank You, Comrade Stalin! Soviet Public Culture from Revolution to Cold War*. Princeton: Princeton University Press, 2000.

------. *When Russia Learned to Read: Literacy and Popular Literature, 1861-1917*. Princeton: Princeton University Press, 1985.

Brooks, Nathan M. "Alexander Butlerov and the Professionalization of Science in Russia." *Russian Review* 57, no. 1 (January 1998): 10-24.

. "The Formation of a Community of Chemists in Russia: 1700-1870." Ph.D. dissertation, Columbia University, 1989.

Bubnov, N. Iu. "Lomonosov: pervye shagi v nauku." In A. A. Zaitseva, ed. *Lomonosov i kniga: sbornik nauchnykh trudov*. Leningrad, 1986, 28-35.

Budilovich, A. S. *M. V. Lomonosov kak naturalist i filolog*. St. Petersburg, 1869.

Burlaka, D. K., et al., eds. *Petr Velikii—pro et contra: lichnost' i deianiia Petra I v otsenke russkikh myslitelei i issledovatelei: antologiia*. St. Petersburg, 2003.

Calinger, Ronald S. "Euler's 'Letters to a Princess of Germany' as an Expression of his Mature Scientific Outlook." *Archive for History of Exact Sciences*, ed. C. Truesdell, vol. 1 (1975/76): 211-33.

------."The Introduction of the Newtonian Natural Philosophy into Russia and Prussia (1725-1772)." Ph.D. dissertation, University of Chicago, 1971.

------. "Leonhard Euler: The First Petersburg Years (1727-1741)." *Historia Mathematica* 23 (1996): 121-166.

Carver, J. Scott. "A Reconsideration of Eighteenth-Century Russia's Contributions to Science." *Canadian-American Slavic Studies* 14, no. 3 (Fall 1980): 389-405.

Chaplin, Joyce E. *The First Scientific American: Benjamin Franklin and the Pursuit of Genius*. New York: Basic Books, 2006.

Chelishchev, P. I. *Puteshestvie po severu Rossii v 1791 godu. Dnevnik P. I. Chelishcheva*. St. Petersburg, 1886.

Chenakal, V. L., ed. *M. V. Lomonosov v portretakh, illiustratsiiakh, dokumentakh*. Moscow–Leningrad, 1965.

------, ed. *Muzei M. V. Lomonosova v Leningrade*. Leningrad, 1967.

Cherniavsky, Michael. *Tsar and People: Studies in Russian Myths*, 2nd ed. New York: Random House, 1969.

Chrissidis, Nikolaos, A. "Creating the New Educated Elite: Learning and Faith in Moscow's Slavo-Greco-Latin Academy, 1685-1694." Ph.D. dissertation, Yale University, 2000.

Clark, William, Jan Golinski, and Simon Schaffer, eds. *The Sciences in Enlightened Europe*. Chicago: University of Chicago Press, 1999.

Cohen, I. Bernard. *Benjamin Franklin's Science*. Cambridge, MA: Harvard University Press, 1990.

Collis, Robert. *The Petrine Instauration: Religion, Esotericism and Science at the Court of Peter the Great, 1689-1735*. Turku, Finland: Turku University, 2007.

Cracraft, James. *The Petrine Revolution in Russian Architecture*. Chicago: University of Chicago Press, 1988.

------. *The Petrine Revolution in Russian Culture*. Cambridge, MA: Harvard University Press, 2004.

------. *The Petrine Revolution in Russian Imagery*. Chicago: University of Chicago Press, 1997.

------, ed. *Peter the Great Transforms Russia*, 3rd ed. Lexington, MA: D.C. Heath and Co., 1991.

Dahlmann, Dittmar, and Galina Smagina, eds. *G. F. Miller i russkaia kul'tura*. St. Petersburg, 2007.

Danilevskii, V. V. *Lomonosov na Ukraine*. Leningrad, 1954.

Danilov, V. V. "Dedushka russkikh istoricheskikh zhurnalov ('Otechestvennye zapiski' P. P. Svin'ina)." *Istoricheskii vestnik: istoriko-literaturnyi zhurnal* 141 (1915): 109-29.

Dashkova, Ekaterina. *The Memoirs of Princess Dashkova*, edited and translated by Kyril Fitzlyon, introduction by Jehanne M. Gheith, afterword by A. Woronzoff-Dashkoff. Durham, NC: Duke University Press, 1995.

David-Fox, Michael. "Symbiosis to Synthesis: The Communist Academy and the Bolshevization of the Russian Academy of Sciences, 1918-1929." *Jahrbücher für Geschichte Osteuropas* 46, no. 2 (1998): 219-43.

Dear, Peter. *Revolutionizing the Sciences: European Knowledge and Its Ambitions, 1500-1700*. Princeton: Princeton University Press, 2001.

Deborin, A. M., ed. *Leonard Eiler, 1707-1783: sbornik statei i materialov k 150-letiiu so dnia smerti*. Moscow-Leningrad, 1935.

Debreczeny, Paul. *Social Functions of Literature: Alexander Pushkin and Russian Culture*. Stanford: Stanford University Press, 1997.

Delbourgo, James. *A Most Amazing Scene of Wonders: Electricity and Enlightenment in Early America*. Cambridge, MA: Harvard University Press, 2006.

Dobrenko, Evgeny. *Stalinist Cinema and the Production of History: Museum of the Revolution*. New Haven: Yale University Press, 2008.

Dovlatov, Sergei. *Chemodan*. Tenafly, NJ: Ermitazh, 1986.

Dvoichenko-Markov, Eufrosina. "The American Philosophical Society and Early Russian-American Relations." *Proceedings of the American Philosophical Society* 94, no. 6 (1950): 549-610.

------. [Dvoichenko-Markoff, Eufrosina]. "Benjamin Franklin, the American Philosophical Society, and the Russian Academy of Science." *Proceedings of the American Philosophical Society* 91, no. 3 (1947): 250-57.

Efremov, P. A., ed. *Materialy dlia istorii russkoi literatury*. St. Petersburg, 1867.

Egorov, B. F. "Lomonosovskii iubilei 1865 g." In *M. V. Lomonosov i russkaia kul'tura: tezisy dokladov konferentsii, posviashchennoi 275-letiiu so dnia rozhdeniia M. V. Lomonosova (28-29 noiabria 1986 g.)*. Tartu, 1986, 56-59.

Eliseev, A. A. *G. V. Rikhman*. Moscow, 1975.

Esakov, V. D., ed. *Akademiia nauk v resheniiakh Politbiuro TsK RKP(b)-VKP(b): 1922-1952*. Moscow, 2000.

Esipov, V. M. *Pushkin v zerkale mifov*. Moscow, 2006.

Evdokimova, Svetlana. *Pushkin's Historical Imagination*. New Haven: Yale University Press, 1999.

Fara, Patricia. *An Entertainment for Angels: Electricity in the Enlightenment*. Cambridge: Icon Books, 2002.

------. *Newton: The Making of Genius*. London: Macmillan, 2002.

Feingold, Mordechai. *The Newtonian Moment: Isaac Newton and the Making of Modern Culture*. Oxford: Oxford University Press, 2004.

Fellmann, Emil A. *Leonhard Euler*. Hamburg: Rowohlt Taschenbuch, 1995.

Figurovskii, N. A., ed. *Istoriia estestvoznaniia v Rossii*. Vol. 1, part 2. Moscow-Leningrad, 1957.

Franklin, Benjamin. *The Papers of Benjamin Franklin*, vol. 12, *January 1 through December 31, 1765*, edited by Leonard W. Laberee. New Haven: Yale University Press, 1968.

Freeze, Gregory L. "The *Soslovie* (Estate) Paradigm and Russian Social History." *American Historical Review* 91, no. 1 (February 1986): 11-36.

Fuss, Nicolas. "Eloge de Monsieur Léonard Euler, lu à l'Académie impériale des sciences de S.-Pétersbourg dans son assemblée du 23 octobre 1783 par M. Nicolas Fuss." *Nova Acta Academiae scientiarum imperialis Petropolitanae* 1 (1783): 159-212.

Gasiorowska, Xenia. *The Image of Peter the Great in Russian Fiction*. Madison, WI: University of Wisconsin Press, 1979.

Gasparov, Boris, Robert P. Hughes, and Irina Paperno, eds. *Cultural Mythologies of Russian Modernism: From the Golden Age to the Silver Age*. Berkeley, University of California Press, 1992.

Ger'e, Vladimir (W. Guerrier). *Leibniz in seinen Beziehungen zu Russland und Peter dem Grossen*. St. Petersburg, 1873.

Gillis, John R., ed. *Commemorations: The Politics of National Identity*. Princeton: Princeton University Press, 1994.

Gleason, Walter J. *Moral Idealists, Bureaucracy, and Catherine the Great.* New Brunswick, NJ: Rutgers University Press, 1981.

Glinka, M. E. *M. V. Lomonosov (opyt ikonografii).* Moscow-Leningrad, 1961.

Golinski, Jan. "Humphry Davy: The Experimental Self." *Eighteenth-Century Studies* 45, no. 1 (2011): 15-28.

------. *Science as Public Culture: Chemistry and Enlightenment in Britain, 1760-1820.* Cambridge: Cambridge University Press, 2003.

Golitsyn, F. N., Kniaz'. "Zhizn' ober-kamergera Ivana Ivanovicha Shuvalova." *Moskvitianin* 6 (1853): 87-98.

Goldin, V. I., ed. *Mikhail Lomonosov: uchenyi—entsiklopedist, poet, khudozhnik, radetel' prosveshcheniia.* Moscow, 2011.

Golubtsov, N. A., ed. *Lomonosovskii sbornik.* Arkhangel'sk, 1911.

Gordin, Michael D. *A Well-Ordered Thing: Dmitrii Mendeleev and the Shadow of the Periodic Table.* New York: Basic Books, 2004.

------. "The Heidelberg Circle: German Inflections on the Professionalization of Russian Chemistry in the 1860s." *Intelligentsia Science: The Russian Century, 1860-1960, OSIRIS 23*, edited by Michael D. Gordin, Karl Hall, and Alexei Kojevnikov, eds. (2008): 23-49.

------. "The Importance of Being Earnest: The Early St. Petersburg Academy of Sciences." *ISIS* 91, no. 1 (March 2000): 1-31.

Gorelik, Gennadii. *Sovetskaia zhzin' L'va Landau.* Moscow, 2008.

Graham, Loren R. "The Birth, Withering, and Rebirth of Russian History of Science." *Kritika: Explorations in Russian and Eurasian History* 2, no. 2 (Spring 2001): 329-40.

------. *The Soviet Academy of Sciences and the Communist Party, 1927-1932.* Princeton: Princeton University Press, 1967.

Greenblatt, Stephen. *Renaissance Self-Fashioning: From More to Shakespeare.* Chicago: University of Chicago Press, 1980.

Greenfeld, Liah. *Nationalism: Five Roads to Modernity.* Cambridge, MA: Harvard University Press, 1992.

Griffiths, David. M. "In Search of Enlightenment: Recent Soviet Interpretation of Eighteenth-Century Russian Intellectual History." *Canadian-American Slavic Studies* 16, nos. 3-4 (Fall-Winter 1982): 317-56.

Gukovskii, G. A. *Russkaia poeziia XVIII veka.* Leningrad, 1927.

Guseva, N. Iu., ed. *M. V. Lomonosov i elizavetinskoe vremia : Katalog vystavki.* St. Petersburg, 2011.

Hall, A. Rupert. *Isaac Newton: Eighteenth-Century Perspectives*. Oxford: Oxford University Press, 1999.

Hankins, Thomas L. "In Defense of Biography: The Use of Biography in the History of Science." *History of Science* 17, no. 35 (March 1979): 1-16.

Haynes, Rosalynn D. *From Faust to Strangelove: Representations of the Scientist in Western Literature*. Baltimore: Johns Hopkins University Press, 1994.

Heilbron, John L. *Electricity in the 17th and 18th Centuries: A Study of Early Modern Physics*. Berkeley: University of California Press, 1979.

------. "Galvani, Volta, and the Uses of Centennials." In *Luigi Galvani International Workshop: Proceedings, Bologna, 9 October 1998*, edited by Marco Bresadola and Guiliano Pancaldi. Bologna: University of Bologna, 1999, 17-32.

Higgitt, Rebekah. *Recreating Newton: Newtonian Biography and the Making of Nineteenth-Century History of Science*. London: Pickering & Chatto, 2007.

Hoffmann, Peter. *Gerhard Friedrich Müller (1705-83): Historiker, Geograph, Archivar im Dienste Russlands*. Frankfurt am Main: Peter Lang, 2005.

------. *Michail Vasil'evic Lomonosov (1711-1765): Ein Enzyklopädist im Zeitalter der Aufklärung*. Frankfurt am Main: Peter Lang, 2011.

Holloway, David. *Stalin and the Bomb: The Soviet Union and Atomic Energy, 1939-1956*. New Haven: Yale University Press, 1994.

Home, R. W. "Leonhard Euler's 'Anti-Newtonian' Theory of Light." *Annals of Science* 45, no. 5 (September 1988): 521-33.

------. "Science as a Career in Eighteenth-Century Russia: The Case of F. U. T. Aepinus." *Slavonic and East European Review* 51, no. 122 (January 1973): 75-94.

Huang, Nian-Sheng. *Benjamin Franklin in American Thought and Culture, 1790-1990*. Philadelphia: American Philosophical Society, 1994.

Hufbauer, Karl. *The Formation of the German Chemical Community (1725-1795)*. Berkeley: University of California Press, 1982.

Hughes, Lindsey. *Peter the Great: A Biography*. New Haven: Yale University Press, 2002.

------. *Russia in the Age of Peter the Great*. New Haven: Yale University Press, 1998.

Hunter, Michael, ed. *Robert Boyle by Himself and His Friends, with a Fragment of William Wotton's Lost 'Life of Boyle.'* London: William Pickering, 1994.

Huntington, W. Chapin. "Michael Lomonosov and Benjamin Franklin: Two Self-Made Men of the Eighteenth Century." *Russian Review* 18, no. 4 (October 1959): 294-306.

Istoriia Moskovskogo universiteta. Vol. 1. Moscow, 1955.

Iushkevich, A. P. "Eiler i russkaia matematika v XVIII v." *Trudy Instituta istorii estestvoznaniia* 3 (1949): 45-116.

------. *Istoriia matematiki v Rossii do 1917 goda.* Moscow, 1968.

------, and Eduard Winter, eds. *Die Berliner und die Petersburger Akademie der Wissenschaften im Briefwechsel Leonhard Eulers.* 3 volumes. Berlin: Akademie-Verlag, 1959 -1976.

Jones, W. Gareth. "Biography in Eighteenth-Century Russia." *Oxford Slavonic Papers* 22 (1989): 58-80.

------. *Nikolay Nokivov: Enlightener of Russia.* Cambridge: Cambridge University Press, 1984.

Joravsky, David. "The Perpetual Province: 'Ever Climbing up the Climbing Wave'." *Russian Review* 57, no. 1 (January 1998): 1-9.

------. *Soviet Marxism and Natural Science, 1917-1932.* New York: Columbia University Press, 1961.

------. "Soviet Views on the History of Science." *ISIS* 46, no. 143 (1955): 3-13.

Josephson, Paul. "Stalinism and Science: Physics and Philosophical Disputes in the USSR, 1930-1955." In *Academia in Upheaval: Origins, Transfers, and Transformations of the Communist Academic Regime in Russia and East Central Europe,* edited by Michael David-Fox and György Péteri, 105-38. Westport, T: Bergin & Garvey, 2000.

Kahn, Andrew. "Self and Sensibility in Radishchev's *Puteshestvie iz Peterburga v Moskvu*: Dialogism and the Moral Spectator." *Oxford Slavonic Papers* 30 (1997): 40-66.

Kapitsa, P. L. "Lomonosov and World Science." In *Collected Papers of P. L. Kapitza,* vol. 3, edited by D. Ter Haar, 168-84. Oxford: Pergamon Press, 1967.

------. "Nauchnaia deiatel'nost' V. Franklina." *Vestnik Akademii nauk SSSR* 2 (1956): 65-75.

Karamzin, N. M. *Izbrannye sochineniia,* vol. 2. Moscow-Leningrad, 1964.

Karpeev, E. P. *Russkaia kul'tura i Lomonosov.* St. Petersburg, 2005.

------. "'Se chelovek...'(zametki k psikhologicheskomu portretu M. V. Lomonosova)." *Voprosy istorii estestvoznaniia i tekhniki* 1 (1999): 106-21.

Kelly, Catriona. *Comrade Pavlik: The Rise and Fall of a Soviet Boy Hero*. London: Granta, 2005.

Khartanovich, M. F. *Uchenoe soslovie Rossii: Imperatorskaia Akademiia nauk vtoroi chetverti XIX v*. St. Petersburg, 1999.

Khartanovich, M. F., and N. P. Kopaneva, eds. *Mikhail Vasil'evich Lomonosov. K 300-letiiu so dnia rozhdeniia: po materialam Muzeia M. V. Lomonosova*. St. Petersburg, 2011.

Klein, Joachim. *Puti kul'turnogo importa: Trudy po russkoi literature XVIII veka*. Moscow, 2005.

------. *Russkaia literatura v XVIII veka*. Moscow, 2010.

Knight, David, and Helge Kragh, eds. *The Making of the Chemist: The Social History of Chemistry in Europe, 1789-1914*. Cambridge: Cambridge University Press, 1998.

Kobeko, D. F. "Uchenik Vol'tera graf Andrei Petrovich Shuvalov." *Russkii Arkhiv*, Book 3 (1881): 241-90.

Kojevnikov, Alexei B. *Stalin's Great Science: The Times and Adventures of Soviet Physicists*. London: Imperial College Press, 2004.

Kopaneva, N. P., and S. B. Koreneva, eds. *G. V. Leibnits i Rossiia*. St. Petersburg, 1998.

Kopelevich, Iu. Kh. *Osnovanie Peterburgskoi Adakemii nauk*. Leningrad, 1977.

------. *Vozniknovenie nauchnykh akademii: seredina XVII – seredina XVIII v*. Leningrad, 1974.

Koreakova, V. "Verevkin, Mikhail Ivanovich." In *Russkii biograficheskii slovar'*. Vol. 3A (Petrograd, 1916), 582-85.

Korshin, Paul J. "The Development of Intellectual Biography in the Eighteenth Century." *Journal of English and Germanic Philology* 73, no. 4 (October 1974): 513-23.

Kovalevskii M. M. "Moskovskii universitet v kontse 70-ikh i nachale 80-ikh godov proshlogo veka (lichnye vospominaniia)," *Vestnik Evropy: zhurnal nauki—politiki—literatury* (May 1910): 178-221.

Koyré, Alexandre. *From the Closed World to the Infinite Universe*. Baltimore: Johns Hopkins University Press, 1957.

Kragh, Helge. *An Introduction to the Historiography of Science*. Cambridge: Cambridge University Press, 1987.

Krementsov, Nikolai. *Stalinist Science*. Princeton: Princeton University Press, 1997.

Kulakova, I. P. *Universitetskoe prostranstvo i ego obitateli: Moskovskii universitet v istoriko-kul'turnoi srede XVIII veka.* Moscow, 2006.

Kulakova, L. I. "Poeziia M. N. Murav'eva." In L. I. Kulakova, ed. *M. N. Murav'ev: stikhotvoreniia,* edited by L. I. Kulakova. Leningrad, 1967, 5-49.

Kuliabko, E. S. "Neizvestnoe pis'mo I. I. Shuvalova k M. V. Lomonosovu." *XVIII vek* 7 (1966): 99-105.

Kunik, A., ed. *Sbornik materialov dlia istorii Imperatorskoi Akademii nauk v XVIII veke,* 2 parts. St. Petersburg, 1865.

Kurilov, A. S., ed. *Lomonosov i russkaia literatura.* Moscow, 1987.

Kuznetsov, B. G. *Tvorcheskii put' Lomonosova.* Moscow, 1961.

Kypriianova. T. G. "Novye arkhivnye svedeniia po istorii sozdaniia 'Arifmetiki' L. Magnitskogo." In ed. *Estestvennonauchnye predstavleniia Drevnei Rusi,* edited by P. A. Simonov, 279-82. Moscow, 1988.

Lamanskii, V. I. *Lomonosov i Peterburgskaia Akademiia nauk. Materialy k stoletiiu pamiati ego, 1765-1865 goda, aprelia 4-go dnia.* Moscow, 1865.

------. *Mikhail Vasil'evich Lomonosov: biograficheskii ocherk.* Reprint, St. Petersburg, 1883.

Latour, Bruno. *The Pasteurization of France.* Translated by Alan Sheridan and John Law. Cambridge, MA: Harvard University Press, 1988.

Lavrent'ev, A. V. *Liudi i veshchi. Pamiatniki russkoi istorii i kul'tury XVI-XVIII vv., ikh sozdateli i vladel'tsy.* Moscow, 1997.

Lavrent'ev, M. A., A. P. Iushkevich, and A. T. Grigor'ian, eds. *Leonard Eiler: sbornik statei v chest' 250-letiia so dnia rozhdeniia, predstavlennykh Akademii nauk SSSR.* Moscow, 1958.

Lebedev, Evgenii. *Lomonosov.* Moscow, 1990.

Leckey, Colum. "What is *Prosveshchenie*? Nikolai Novikov's *Historical Dictionary of Russian Writers* Revisited." *Russian History* 37 (2010): 360-377.

Leicester, Henry M. "Boyle, Lomonosov, Lavoisier, and the Corpuscular Theory of Matter." *ISIS* 58, no. 192 (Summer 1967): 240-45.

------. "Znakomstvo uchenykh Severnoi Ameriki kolonial'nogo perioda s rabotami M. V. Lomonosova i Peterburgskoi Akademii nauk." *Voprosy istorii estestvoznaniia i tekhniki* 12 (1962): 142-47.

Levin, Iu. D. *Vospriiatie angliiskoi literatury v Rossii.* Leningrad, 1990.

Levitt, Marcus C. "An Antidote to Nervous Juice: Catherine the Great's Debate with Chappe d'Auteroche over Russian Culture." *Eighteenth-Century Studies* 32, no. 1 (1998): 49-63.

------. *Russian Literary Politics and the Pushkin Celebration of 1880*. Ithaca, NY: Cornell University Press, 1989.

------. *The Visual Dominant in Eighteenth-Century Russia*. DeKalb, IL: Northern Illinois University Press, 2012.

------, ed. *Dictionary of Literary Biography*. Vol. 150, *Early Modern Russian Writers: Late Seventeenth and Eighteenth Centuries*. Detroit: Gale Research Company, 1995.

Levshin, L.V. *Sergei Ivanovich Vavilov, 1891-1951*. Moscow, 2003.

Liechtenhan, Francine-Dominique. "Jacob von Stählin, académicien et courtesan." *Cahiers du monde russe* 43, no 2-3 (2002): 321-32.

Liubimov, N. A. "Lomonosov kak fizik." In *V vospominanie 12-go ianvaria 1855 goda. Ucheno-literaturnye stat'i professorov i prepodavatelei Moskovskogo universiteta*, 3-35. Moscow, 1855.

------. *Zhizn' i trudy Lomonosova: s prilozheniem ego portreta*. Moscow, 1872.

"Liubimov, Nikolai Alekseevich." In *Entsiklopedicheskii slovar'* (*Brockhaus-Efron*), vol. 18 (St. Petersburg, 1896), 209.

Livingstone, David N. *Putting Science in its Place: Geographies of Scientific Knowledge*. Chicago: University of Chicago Press, 2003.

Lomonosov, M. V. *Mikhail Vasil'evich Lomonosov on the Corpuscular Theory*, translated and with an introduction by Henry M. Leicester. Cambridge, MA: Harvard University Press, 1970.

------. *Mikhail Vasil'evich Lomonosov: perepiska, 1737-1765*. Moscow, 2010.

------. *Polnoe sobranie sochinenii Mikhaila Vasil'evicha Lomonosova, s priobshcheniem zhizni sochinitelia i s pribavleniem mnogikh ego nigde eshche ne napechatannykh tvorenii*, 6 volumes. St. Petersburg, 1784-87.

------. *Polnoe sobranie sochinenii*, 11 volumes. Moscow-Leningrad, 1950-83.

------. *Sobranie raznykh sochinenii v stikhakh i v proze*, 2 volumes. Moscow, 1757-65.

------. *Sobranie raznykh sochinenii v stikhakh i v proze*, 3 volumes. Moscow, 1778.

------. *Sobranie sochinenii izvestneishikh russkikh pisatelei*, no. 1, *Izbrannyia sochineniia M. V. Lomonosova, s ego portretom, biografieiu, spiskom s pocherka i s izlozheniem soderzhaniia statei o Lomonosove, napechatannykh v raznykh periodicheskikh i dr. izdaniiakh*, edited by P. Perevlesskii. Moscow, 1846.

------. *Sochineniia M. V. Lomonosova*, 8 volumes. St. Petersburg-Moscow-Leningrad, 1891-1948.

Lomonosov: sbornik statei i materialov, 10 volumes. Moscow-Leningrad-St. Petersburg, 1940-2011.

Lomonosovskii sbornik. St. Petersburg, 1911.

Lomonosovskii sbornik: materialy dlia istorii razvitiia khimii v Rossii. Moscow, 1901.

Lotman, Iu. M. *O russkoi literature: stat'i i issledovaniia (1958-1993)*. St. Petersburg, 1997.

Lotman, Iu. M. and B. A. Uspenskii, "K semioticheskoi tipologii russkoi kul'tury XVIII veka." In *Iz istorii russkoi kul'tury*, edited by A. D. Koshelov, vol. 4 (XVIII-nachalo XIX veka), 425-47. Moscow, 1996.

------. "Myth-Name-Culture." In *Soviet Semiotics: An Anthology*, edited and translated by Daniel P. Lucid. Baltimore: Johns Hopkins University Press, 1988, 233-52.

------. *The Semiotics of Russian Culture*. Edited by Ann Shukman. Translated by N. F. C. Owen. Ann Arbor, MI: Department of Slavic Languages and Literatures, University of Michigan, 1984, 3-35.

Lukina, T. A. "Ekspeditsii akademika Lepekhina v XVIII v." *Trudy Instituta istorii estestvoznaniia i tekhniki* 41 (1961): 324- 52.

Luk'ianov, P. M. "A. N. Radishchev i khimiia." *Trudy Instituta istorii estestvoznaniia i tekhniki* 2 (1954): 158-67.

L'vovich-Kostritsa, A. I. *M. V. Lomonosov, ego zhizn', nauchnaia, literaturnaia i obshchestvennaia deiatel'nost': biograficheskii ocherk*. St. Petersburg, 1892.

Lystsov, V. P. *M. V. Lomonosov v russkoi istoriografii 1750-1850-kh godov*. Voronezh, 1983.

------. *M. V. Lomonosov v russkoi istoriografii 1860-1870-kh godov*. Voronezh, 1992.

------. *Zhizn' i deiatel'nost' M. V. Lomonosova v osveshchenii P. P. Pekarskogo*. Voronezh, 1993.

Maggs, Barbara Widenor. "Firework Art and Literature: Eighteenth-Century Pyrotechnical Tradition in Russia and Western Europe." *Slavonic and East European Review* 54, no. 1 (January 1976): 24-40.

Makarov, V. K. *Khudozhestvennoe nasledie M. V. Lomonosova: mozaiki*. Moscow-Leningrad, 1950.

Makogonenko, G. P. *Radishchev i ego vremia*. Moscow, 1956.

------, ed. *Pis'ma russkikh pisatelei XVIII veka*. Leningrad, 1980.

Manuel, Frank. E. *A Portrait of Isaac Newton*. New York: De Capo, 1968.

Marker, Gary. "Standing in St. Petersburg Looking West, Or, Is Backwardeness All There Is?" *Republic of Letters: A Journal for the Study of Knowledge, Politics, And the Arts* 1, no. 1 (May 2009).

Martynov, G. G., ed. *Mikhail Lomonosov: glazami sovremennikov*. Moscow, 2011.

------, ed. *Mikhail Vasil'evich Lomonosov: perepiska* 1737-1765. Moscow, 2010.

Martynov, I. F. "'Opyt istoricheskogo slovaria o rossiiskikh pisateliakh' N. I. Novikova i literaturnaia polemika 60-70-kh godov XVIII veka." *Russkaia literatura*, no. 3 (1968): 184-91.

Mashkova, M. V. *P. P. Pekarskii (1827-1872): kratkii ocherk zhizni i deiatel'nosti*. Moscow, 1957.

McClellan, James E. III. *Science Reorganized: Scientific Societies in the Eighteenth Century*. New York: Columbia University Press, 1985.

McClelland, James C. *Autocrats and Academics*. Chicago: University of Chicago Press, 1979.

McConnell, Allen. *A Russian Philosophe: Alexander Radishchev*. The Hague: M. Nijhoff, 1964.

Mel'nikov, P. I. *Opisanie prazdnestva, byvshago v S.-Peterburge 6-9 aprelia 1865 g. po sluchaiu stoletniago iubileia Lomonosova*. St. Petersburg, 1865.

Menshutkin, B. N. *Lomonosov kak estestvoispytatel'*. St. Petersburg, 1911.

------. *Lomonosov kak fiziko-khimik: k istorii khimii v Rossii*. St. Petersburg, 1904.

------. *Mikhailo Vasil'evich Lomonosov: zhizneopisanie*. St. Petersburg, 1911.

------. *Russia's Lomonosov: Chemist, Courtier, Physicist, Poet*. Translated by Jeannette Eyre Thal and Edward J. Webster. Princeton: Princeton University Press, 1952.

------. "Trudy M. V. Lomonosova po fizike i khimii." In *Trudy Lomonosova v oblasti estestvenno-istoricheskikh nauk*. St. Petersburg, 1911, 1-103.

------. *Trudy M. V. Lomonosova po fizike i khimii*. Moscow-Leningrad, 1936.

------. *Zhizneopisanie Mikhaila Vasil'evicha Lomonosova*, 2nd ed. Moscow-Leningrad, 1937.

------. *Zhizneopisanie Mikhaila Vasil'evicha Lomonosova*, 3rd ed. Edited by P. N. Berkov, S. I. Vavilov, and L. B. Modzalevskii. Moscow-Leningrad, 1947.

------. *Zhizn' i deiatelnost' Nikolaia Aleksandrovicha Menshutkina*. St. Petersburg, 1908.

Mikhailo Lomonosov (film). Directed by Mikhail Shapiro. Leningrad, 1955.

Mikhailo Lomonosov (film/television series). Directed by Aleksandr Proshkin. Moscow, 1984-86.

Minaeva, O. D. <<*Otechestva ymnozhit' slavu*>>: *biografiia M. V. Lomonosova*. Moscow, 2011.

Molok, Iurii. *Pushkin v 1937 godu: materialy i issledovaniia po ikonografii*. Moscow, 2000.

Moiseeva, G. N. *Lomonosov i drevnerusskaia literatura*. Leningrad, 1971.

Moriakov, V. I. "A. N. Radishchev o M. V. Lomonosove." *Vestnik Moskovskogo universiteta*, series 8, history, no. 4 (July-August 1986): 34-43.

Morozov, A. A. *M. V. Lomonosov: put' k zrelosti, 1711-1741*. Moscow-Leningrad, 1962.

------. *Mikhail Vasil'evich Lomonosov*, 5th ed. Moscow, 1965.

------. *Rodina Lomonosova*. Arkhangel'sk, 1975.

Mühlpfordt, Günter. "Deutsch-russische Wissenschaftsbeziehungen in der Zeit der Aufklärung. Christian Wolff und die Gründung der Petersburger Akademie der Wissenschaften." In *450 Jahre Martin-Luther-Universität Halle-Wittenberg*, vol. 2, edited by Leo Stern. Halle 1952, 169-97.

Mumentaler, Rudol'f. *Shveitsarskie uchenye v Sankt-Peterburgskoi akademii nauk XVIII vek*. St. Petersburg, 2009.

Murav'ev, M. N. *Institutiones Rhetoricae. A Treatise of a Russian Sentimentalist*, edited and with an introduction by Andrew Kahn. Oxford: Willem A. Meeuws, 1995.

------. *M. N. Murav'ev: stikhotvoreniia*. Edited by L. I. Kulakova. Leningrad, 1967.

------. *Pokhval'noe slovo Mikhaile Vasil'evichu Lomonosovu pisal leib-gvardii Izmailovskago polku kaptenarmus Mikhailo Murav'ev*. St. Petersburg, 1774.

------. *Polnoe sobranie sochinenii M. N. Murav'eva*, 3 vols. St. Petersburg, 1819-20.

------. *Sochineniia M. N. Murav'eva*, 2 vols. St. Petersburg, 1847.

------. "Zaslugi Lomonosova v uchenosti." In *M. N. Murav'ev, Opyty istorii, pis'men i nravoucheniia* 32-39. St. Petersburg, 1796.

"Murav'ev (Mikhail Nikitich)—obshestvennyi deiatel' i pisatel' (1757-1800)." In *Entsiklopedicheskii slovar' (Brockhaus-Efron)*, vol. 20, 189-190. St. Petersburg, 1897.

M. V. Lomonosov v knizhnoi kul'ture Rossii (Moscow, 2011).

Newman, William R. *Atoms and Alchemy: Chymistry & the Experimental Origins of the Scientific Revolution*. Chicago: University of Chicago Press, 2006.

Nikolaev, S. I., ed. *Peter I v russkoi literature XVIII veke: teksty i kommentarii*. St. Petersburg, 2007.

Nora, Pierre, ed. *Realms of Memory: The Construction of the French Past*, 3 vols, translated by Arthur Goldhammer. New York: Columbia University Press, 1996-98.

Novik, V. K. "Akademik Frants Epinus (1724-1802): kratkaia biograficheskaia khronika." *Voprosy istorii estestvoznaniia i tekhniki* 4 (1999): 4-35.

Novikov, N., ed. *Opyt istoricheskago slovaria o rossiiskikh pisateliakh*. St. Petersburg, 1772. Reprint, Leningrad, 1987.

Nye, Mary Jo. "Scientific Biography: History of Science by Another Means?" *ISIS* 97, no. 2 (June 2006): 322-29.

Okenfuss, Max J. *The Rise and Fall of Latin Humanism in Early Modern Russia: Pagan Authors, Ukrainians, and the Resiliency of Muscovy*. Leiden: E. J. Brill, 1995.

Orel, V. M., and Galina Smagina, eds. *Komissiia po istorii znanii 1921-1932 gg. Iz istorii organizatsii istoriko-nauchnykh issledovanii v Akademii nauk: sbornik dokumentov*. St. Petersburg, 2003.

Osipov, Iu. S. *Akademiia nauk v istorii Rossiiskogo gosudarstva*. Moscow, 1999.

Ospovat, Kirill. "Lomonosov i 'pismo o pol'ze stekla': poeziia i nauka pri dvore Elizavety petrovny." *Novoe literaturnoe obozrenie* 87 (2007): 148-83.

------. "Mikhail Lomonosov Writes to his Patron: Professional Ethos, Literary Rhetoric and Social Ambition." *Jahrbücher für Geschichte Osteuropas* 59, no. 2 (2011): 240-66.

Ostrovitianov, K. V., ed. *Istoriia Akademii nauk SSSR*, 2 vols. Moscow-Leningrad, 1958-64.

Outram, Dorinda. *Georges Cuvier: Science, Vocation and Authority in Post-Revolutionary France*. Manchester: Manchester University Press, 1984.

------. "The Language of Natural Power: The 'Eloges' of Georges Cuvier and the Public Language of Nineteenth-Century Science." *History of Science* 16, no. 33 (September 1978): 153-78.

------. "Scientific Biography and the Case of Georges Cuvier: With a Critical Bibliography." *History of Science* 14, no. 24 (June 1976): 101-37.

Pamiati Lomonosova, 6-go aprelia 1865 goda. Khar'kov, 1865.

Pancaldi, Guiliano. *Volta: Science and Culture in the Age of Enlightenment.* Princeton: Princeton University Press, 2003.

Paul, C. B. *Science and Immortality: The Eloges of the Paris Academy of Sciences (1699-1791).* Berkeley: University of California Press, 1980.

Pavlenko, N. I. *Mikhail Pogodin.* Moscow, 2003.

Pavlova, G. E. "Lomonosov v kharakteristikakh i vospominaniiakh sovremennikov." *Voprosy istorii estestvoznaniia i tekhniki* 3 (1986): 59-69.

------. "Proekty illiuminatsii Lomonosova." In *Lomonosov: sbornik statei i materialov,* vol. 4. Moscow-Leningrad, 1960, 219-37.

------. *Stepan Iakovlevich Rumovskii, 1734-1812.* Moscow, 1979.

------, ed. *M. V. Lomonosov v vospominaniiakh i kharakteristikakh sovremennikov.* Moscow-Leningrad, 1962.

Pavlova, G. E., and A. S. Fedorov. *Mikhail Vasil'evich Lomonosov, 1711-1765.* Moscow, 1986.

Pekarskii, P. P. *Dopolnitel'nyia izvestiia dlia biografii Lomonosova.* St. Petersburg, 1865.

------. *Istoriia Imperatorskoi Akademii nauk v Peterburge.* 2 vols. St. Petersburg, 1870-73.

------. *Nauka i literatura v Rossii pri Petre Velikom.* Vol. 1, *Vvedenie v istoriiu prosveshcheniia v Rossii XVIII stoletiia.* St. Petersburg, 1862.

------. "O rechi v pamiat' Lomonosova, proiznesennoi v Akademii nauk doktorom Le-Klerkom," *Zapiski Imperatorskoi Akademii nauk* 10, book 2 (1867): 178-81.

[Perevoshchikov, D. M.?]. "O fizicheskikh sochineniiakh Lomonosova." *Atenei* 2 (January 1829): 109-20.

Perevoshchikov, D. M. "Rassmotrenie Lomonosova razsuzhdeniia: 'o iavleniiakh vozdushnykh, ot eliktricheskoi sily proizkhodiashchikh'." *Teleskop,* no. 4 (1831): 486-513.

------. *Rukovodstvo k opytnoi fizike.* Moscow, 1833.

Petukhov, E. "Mikhail Nikitich Murav'ev: ocherk ego zhizni i deiatel'nosti." *Zhurnal Ministerstva narodnogo prosveshcheniia* 294, section 2 (August 1894): 265-96.

Plamper, Jan. *The Stalin Cult: A Study in the Alchemy of Power.* New Haven: Yale University Press, 2012.

Platt, Kevin M. F. *Terror and Greatness: Ivan and Peter as Russian Myths.* Ithaca, NY: Cornell University Press, 2011.

Pliukhanova, M. B. "'Istoricheskoe' i 'mifologicheskoe' v rannikh biografiiakh Petra I." In *Vtorichnye modeliruiushchie sistemy*, 82-88. Tartu, 1979.

Pogodin, M. P. "Petr Velikii." In M. P. Pogodin, *Istoriko-kriticheskie otryvki*, 333-63. Moscow, 1846.

------. "Vospominanie o Lomonosove." *Moskvitianin* 2 (1855): 1-16.

Pogodin, S. A., and N. M. Raskin. "B. N. Menshutkin kak issledovatel' trudov Lomonosova po khimii i fizike." In *Lomonosov: sbornik statei i materialov*, vol. 6, 245-66. Moscow-Leningrad, 1965.

Pogosian, Elena. *Vostorg russkoi ody i reshenie temy poeta v russkom panegirike 1730-1762 gg.* Tartu, 1997.

Poirier, Jean-Pierre. *Lavoisier: Chemist, Biologist, Economist.* Translated by Rebecca Balinski. Philadelphia: University of Pennsylvania Press, 1996.

Poliarnaia zvezda, izdannaia A. Bestuzhevym i K. Ryleevym. Moscow-Leningrad, 1960.

Polevoi, K. *Mikhail Vasil'evich Lomonosov*, 2 vols. Moscow, 1836 (1st volume); St. Petersburg, 1887 (reprint, 2nd volume).

Pollock, Ethan. *Stalin and the Soviet Science Wars.* Princeton: Princeton University Press, 2006.

Pomper, Philip. "Lomonosov and the Discovery of the Law of the Conversation of Matter in Chemical Transformations." *Journal of the Society for the Study of Alchemy and Early Chemistry (Ambix)* 10, no. 3 (October 1962): 119-27.

Prazdnovanie stoletnei godovshchiny Lomonosova 4 aprelia 1765-1865 g. Imperatorskim Moskovskim universitetom v torzhestvennom sobranii aprelia 11-go dnia. Moscow, 1865.

Prince, Sue Ann, ed. *The Princess & The Patriot: Ekaterina Dashkova, Benjamin Franklin, and the Age of Enlightenment.* Philadelphia: American Philosophical Society, 2006.

Protokoly zasedanii Konferentsii Imperatorskoi Akademii nauk s 1725 po 1803 goda, vol. 2, 1744-1770. St. Petersburg, 1899.

Pushkin, A. S. "O predislovii g-na Lemonte k perevodu basen I. A. Krylova." *Moskovskii telegraf*, part 5, no. 17 (1825): 40-46.

------. *Polnoe sobranie sochinenii*, vols. 11-14. Moscow-Leningrad, 1937-49.

------. *Pushkin on Literature.* Edited and translated by Tatiana Wolff. Evanston, IL: Northwestern University Press, 1998.

Puteshestviia akademika Ivana Lepekhina v 1772 godu, part 4. St. Petersburg, 1805.

Putilov, B. N., ed. *Petr Velikii v predaniiakh, legendakh, anekdotakh, skazkakh, pesniakh*. St. Petersburg, 2000.

Rabique, Charles. *Lettre élèctrique sur la mort de M. Richmann*. Paris, 1753.

Radishchev, A. N. *A Journey from St. Petersburg to Moscow*. Edited by Roderick Page Thaler, translated by Leo Weiner. Cambridge, MA: Harvard University Press, 1958.

------. *Polnoe sobranie sochinenii*, 3 vols. Moscow-Leningrad, 1938-52.

------. *Puteshestvie iz Peterburga v Moskvu. Vol'nost*. Edited by V. A. Zapadov. St. Petersburg, 1992.

Radishchev, P. A. "A. N. Radishchev." *Russkii vestnik* 18, book 1 (December 1858): 395-432.

Radovskii, M. I. *M. V. Lomonosov i Peterburgskaia Akademiia nauk*. Moscow-Leningrad, 1961.

------. *Veniamin Franklin i ego sviazi s Rossiei*. Moscow-Leningrad, 1958.

Raeff, Marc. "Russian Intellectual History and its Historiography." *Forschungen Zur Osteuropäischen Geschichte* 25 (1978): 297-303.

Raikov, B. E. *Ocherki po istorii geliotsentricheskogo mirovozzreniia v Rossii: iz proshlogo russkogo estestvoznaniia*, 2nd ed. Moscow-Leningrad, 1947.

Rainov, T. I. "Russkoe estestvoznanie vtoroi poloviny XVIII v. i Lomonosov." In *Lomonosov: sbornik statei i materialov*, vol. 1. Moscow-Leningrad, 1940, 318-88.

Raskin, N. M. *Khimicheskaia laboratoriia M. V. Lomonosova. Khimia v Peterburgskoi Akademii nauk vo 2-i polovine XVIII v*. Moscow-Leningrad, 1962.

------. *Vasilii Ivanovich Klement'ev—uchenik i laborant M. V. Lomonosova*. Moscow-Leningrad, 1952.

Rebekkini, Damiano. "Russkie istoricheskie romany 30-x godov XIX veka." *Novoe literaturnoe obozrenie* 34 (1998): 416-33.

Reyfman, Irina. *Vasilii Trediakovsky: The Fool of the 'New' Russian Literature*. Stanford: Stanford University Press, 1990.

Riasanovsky, Nicholas V. *The Image of Peter the Great in Russian History and Thought*. New York: Oxford University Press, 1985.

Rice, Adrian. "Augustus De Morgan: Historian of Science." *History of Science* 34, no. 104 (June 1996): 201-40.

Richter, Liselotte. *Leibniz und sein Russlandbild*. Berlin: Deutsche Akademie der Wissenschaften zu Berlin, 1946.

Rikhman. G.-V. *Trudy po fizike*. Moscow, 1956.

Röhling, Hans. "Illustrated Publications on Fireworks and Illuminations in Eighteenth-Century Russia." In A. G. Cross, ed. *Russia and the West in the Eighteenth Century*, edited by A. G. Cross, 94-100. Newtonville, MA: Oriental Research Partners, 1983, 94-100.

Rovinskii, D. A. *Obozrenie ikonopisaniia v Rossii do kontsa XVIII veka. Opisanie feierverkov i illiuminatsii*. Moscow, 1903.

Rupke, Nicolaas A. *Alexander von Humboldt: A Metabiography*. Chicago: University of Chicago Press, 2008.

Rychalovskii, E. E., ed. *Istoriia Moskovskogo universiteta (vtoraia polovina XVIII—nachala XIX veka). Sbornik dokumentov*. Vol. 1: *1754-1755*. Moscow, 2006.

Sandler, Stephanie. *Commemorating Pushkin: Russia's Myth of a National Poet*. Stanford: Stanford University Press, 2004.

Sanktpeterburgskie vedomosti (1746-65).

Scheibert, Peter. "Lomonosov, Christian Wolff und die Universität Marburg." In *Academia Marburgensis: Beiträge zur Geschichte der Philipps-Universität Marburg*, edited by Walter Heinemeyer, Thomas Klein, and Hellmut Seier, 231-40. Marburg, 1977.

Schenck, Frithjof Benjamin. *Aleksandr Nevskij: Heiliger—Fürst—Nationalheld. Eine Erinnerungsfigur im russischen kulturellen Gedächtnis (1263-2000)*. Köln: Böhlau, 2004.

Schönle, Andreas. *Authenticity and Fiction in the Russian Literary Journey, 1790-1840*. Cambridge, MA: Harvard University Press, 2000.

Schulze, Ludmilla. "The Russification of the St. Petersburg Academy of Sciences and Arts in the Eighteenth Century." *British Journal for the History of Science* 18 (1985): 305-35.

Semennikov, V. P. *Knigoizdatel'skaia deiatel'nost' N. I. Novikova i tipograficheskoi kompanii*. Petrograd, 1921.

Serman, Il'ia Z. *Mikhail Lomonosov: Life and Poetry*, translated by Stephany Hoffman. Jerusalem: Hebrew University of Jerusalem, 1988.

------. "<<Slovo o Lomonosove>> i ego mesto v <<Puteshestvii iz Peterburga v Moskvu>>." In *Problemy izucheniia russkoi literatury XVIII veka: Mezhvuzovskii sbornik nauchnykh trudov*, edited by E. I. Annenkova and O. M. Buranok, 222-32. Samara, 2001.

Severgin, V. M. *Slovo pokhval'noe Mikhailu Vasil'evichu Lomonosovu*. St. Petersburg, 1805.

Shank. J. B. *The Newton Wars and the Beginning of the French Enlightenment*. Chicago: University of Chicago Press, 2008.

Shapin, Steven. *A Social History of Truth: Civility and Science in Seventeenth-Century England*. Chicago: University of Chicago Press, 1994.

------. "The Image of the Man of Science." In *The Cambridge History of Science*, vol. 4: *Eighteenth-Century Science*, edited by Roy Porter, 159-83. Cambridge: Cambridge University Press, 2003.

------. *The Scientific Life: A Moral History of a Late Modern Vocation*. Chicago: University of Chicago Press, 2008.

Shapin, Steven, and Simon Schaffer. *Leviathan and the Air Pump: Hobbes, Boyle, and the Experimental Life*. Princeton: Princeton University Press, 1985.

Sheptunova, Z. I. *Istoriograficheskii analiz rabot po istorii khimii v Rossii XVIII—nachalo XX v*. Moscow, 1995.

Shevyrev, Stepan. *Istoriia Imperatorskogo Moskovskogo universiteta, 1755-1855*. Moscow, 1855.

Shmurlo, E. *Petr Velikii v otsenke sovremennikov i potomstva*. St. Petersburg, 1912.

Shortland, Michael, and Richard Yeo, eds. *Telling Lives in Science: Essays on Scientific Biography*. Cambridge: Cambridge University Press, 1996.

Shubinskii, Valerii. *Mikhail Lomonosov: vserossiiskii chelovek*. St. Petersburg, 2006.

"Shuvalov, Ivan Ivanovich." In *Russkii biograficheskii slovar'*, vol. 23 (St. Petersburg, 1911; reprint, New York, 1962), 476-86.

Sivkov, K. V. *Puteshestviia russkikh liudei za granitsu v XVIII veke*. St. Petersburg, 1914.

Slater, Wendy. "The Patriots' Pushkin." *Slavic Review* 58, no. 2 (Summer 1999): 407-27.

Smagina, Galina. *Akademiia nauk i rossiiskaia shkola. Vtoraia polovina XVIII veka*. St. Petersburg, 1996.

------. *Kniaginia i uchenyi: E.R. Dashkova i M.V. Lomonosov*. St. Petersburg, 2011.

------, ed. *Nemtsy v Rossii: tri veka nauchnogo sotrudnichestva*. St. Petersburg, 2003.

Smirnov, S. K. *Istoriia Moskovskoi slaviano-greko-latinskoi akademii.* Moscow, 1855.

Smith, Alexander. "An Early Physical Chemist-M. W. Lomonossoff." *The Journal of the American Chemical Society* 34, no. 2 (February 1912): 109-19.

Smith, G. S. "The Most Proximate West: Russian Poets and the German Academicians, 1728-41." In *Russia and the World of the Eighteenth Century,* edited by R. P. Bartlett, A. G. Cross, and Karen Rasmussen, 360-70. Columbus, Ohio: Slavica Publishers, 1988.

Smolegovskii, A. M., and Iu. I. Solov'ev. *Boris Nikolaevich Menshutkin: Zhizn' i istorik nauki.* Moscow, 1983.

Soboleva, E. V. *Bor'ba za reorganizatsiiu Peterburgskoi Akademii nauk v seredine XIX veka.* Leningrad, 1971.

------. *Organizatsiia nauki v poreformennoi Rossii.* Leningrad, 1983.

Söderqvist, Thomas, ed. *The History and Poetics of Scientific Biography.* Burlington, VT: Ashgate, 2007.

Sokhatskii, P. A. *Slovo na poluvekovoi iubilei Moskovskogo universiteta.* Moscow, 1805.

Sokolov, A. P., ed. *Proekt Lomonosova i ekspeditsiia Chichagova.* St. Petersburg, 1854.

Sokolova, N. V. "Kratkii obzor angliiskoi literatury XVIII-XIX vv. o M. V. Lomonosove." In *Lomonosov: sbornik statei i materialov,* vol. 7, 160-77. Leningrad, 1977.

Solov'ev, Iu. I. *Istoriia khimii v Rossii,* Moscow, 1985.

------. "M. V. Lomonosov v otsenke A. S. Pushkina." *Voprosy istorii estestvoznaniia i tekhniki* no. 4 (1983): 65-69.

Solov'ev, Iu. I., and N. N. Ushakova, *Otrazhenie estestvennonauchnykh trudov M. V. Lomonosova v russkoi literature XVIII i XIX vv.* Moscow, 1961.

Somov, V. A. "N. –G. Leklerk o M. V. Lomonosove." In *Lomonosov: sbornik statei i materialov,* vol. 8, 97-105. Leningrad, 1983.

Sonin, A. S. *Fizicheskii idealizm: istoriia odnoi idelologicheskoi kampanii.* Moscow, 1994.

Sonntag, Otto. "The Motivations of the Scientist: The Self-Image of Albrecht von Haller." *ISIS* 65, no. 228 (September 1974): 336-51.

Staehlin, Jacob von. "Cherty i anekdoty dlia biografii Lomonosova, vziatye s ego sobstvennykh slov Shtelenym." *Moskvitianin* 1 (1850): 1-14.

------. "Konspekt pokhval'nogo slova Lomonosovu," *Moskvitianin* 3 (1853): 22-25.

------. *Originalanekdoten von Peter dem Grossen.* Leipzig, 1785.

------. *Zapiski Iakoba Shtelina ob iziashchnykh iskusstvakh v Rossii,* 2 vols, edited by K. V. Malinovskii. Moscow, 1990.

Stennik, Iu. V. *Ideia "drevnei" i "novoi" Rossii v literature i obshchestvenno-istoricheskoi mysli XVIII—nachala XIX veka.* St. Petersburg, 2004.

------. *Pushkin i russkaia literature XVIII veka.* St. Petersburg, 1995.

------. "Verevkin, Mikhail Ivanovich." In *Slovar' russkikh pisatelei XVIII veka,* no. 1, 148-50. Leningrad, 1988.

Stewart, Larry. *The Rise of Public Science: Rhetoric, Technology, and Natural Philosophy in Newtonian Britain, 1600-1750.* Cambridge: Cambridge University Press, 1993.

Stoletnii iubilei Imperatorskago Moskovskago universiteta. Moscow, 1855.

Sukhomlinov, M. I. *Istoriia Rossiiskoi Akademii.* Vol. 4. St. Petersburg, 1878.

------. "K biografii Lomonosova." In *Izvestiia Otdeleniia russkogo iazyka i slovesnosti Imperatorskoi Akademii nauk* 1, book 4, 779-91. St. Petersburg, 1896.

------. "Lomonosov—student Marburgskogo universiteta." *Russkii vestnik* 31, no. 1 (January 1861): 127-65.

Sumarokov, A. P. *Izbrannye proizvedenniia.* Leningrad, 1957.

Svin'in, P. P. "Izvestie o vnov' otkrytykh rukopisiakh Lomonosova." *Otechestvennye zapiski* 31, no. 89 (September 1827): 489-94.

------. "Potomki i sovremenniki Lomonosova." *Biblioteka dlia chteniia* 2 (1834): 213-20.

Swoboda, Marina, and William Benton Whisenhunt. *A Russian Paints America: The Travels of Pavel P. Svin'in, 1811-1813.* Montreal: McGill-Queen's University Press, 2008.

Tartakovskii, A. G. "A. S. Pushkin i A. N. Radishchev." *Otechestvennaia istoriia* 1-2 (January-February 1999): 64-90; (March-April 1999): 142-70.

------. *Russkaia memuaristika XVIII-pervoi poloviny XIX v. Ot rukopisi k knige.* Moscow, 1991.

Tatarintsev, A. G. "'Slovo o Lomonosove' A. N. Radishcheva. (K probleme tvorcheskoi istorii 'Puteshestviia')." In *Voprosy russkoi i zarubezhnoi literatury,* 17-36. Perm', 1974.

Terras, Victor. "Some Observations on Pushkin's Image in Russian Literature." *Russian Literature* 14 (1983): 296-316.

Tertz, Abram (Andrei Sinyavsky). *Strolls with Pushkin.* Translated by Catherine Theimer Nepomnyashchy and Slava I. Yastremski. New Haven: Yale University Press, 1993.

Thaden, Edward C. *The Rise of Historicism in Russia*. New York: Peter Lang, 1999.

Theerman, Paul. "Unaccustomed Role: The Scientist as Historical Biographer—Two Nineteenth-Century Portrayals of Newton." *Biography* 8, no. 2 (Spring 1985): 145-62.

Tiulichev, D. V. *Knigoizdatel'skaia deiatel'nost' Peterburgskoi Akademii nauk i M. V. Lomonosov*. Leningrad, 1988.

Todd, Williams Mills III. *Fiction and Society in the Age of Pushkin: Ideology, Institutions, and Narrative*. Cambridge, MA: Harvard University Press, 1986.

Tolstoi, D. A. *Akademicheskii universitet v XVIII stoletii po rukopisnym dokumentam Arkhiva Akademii nauk*. St. Petersburg, 1885.

Tolz, Vera. *Russian Academicians and the Revolution: Combining Professionalism and Politics*. New York: St. Martin's Press, 1997.

Toporov, V. N. *Iz istorii russkoi literatury*, vol. 2, *Russkaia literatura vtoroi poloviny XVIII veka: issledovaniia, materialy, publikatsii*. M. N. Murav'ev: vvedenie v *tvorcheskoe nasledie*, books 1-3. Moscow, 2001-2007.

Trigos, Ludmilla A. *The Decembrist Myth in Russian Culture*. New York: Palgrave Macmillan, 2009.

Tropp, E. A., and G. I. Smagina, eds. *Akademiia nauk v istorii kul'tury Rossii XVIII-XX vekov*. St. Petersburg, 2010.

------. *Akademicheskaia nauka v Sankt-Peterburge v XVIII-XX vekakh: istoricheskie ocherki*. St. Petersburg, 2003.

Tsverva, G. K. *Georg Vil'gel'm Rikhman (1711-1753)*. Leningrad, 1977.

Tumarkin, Nina. *Lenin Lives! The Lenin Cult in Soviet Russia*. Enlarged Edition, Cambridge, MA: Harvard University Press, 1997.

Tunkina, I. V. "Lavrentii Lavrent'evich Bliumentrost." In *Vo glave pervenstvuiushchego uchenogo sosloviia Rossii: ocherki zhizni i deiatel'nosti prezidentov Imperatorskoi Sankt-Peterburgskoi Akademii nauk. 1725-1917gg*, edited by V. S. Solovev. St. Petersburg, 2000, 13-28.

Uraniia. Karmannaia knizhka na 1826 god dlia liubitel'nits i liubitelei russkoi slovesnosti. Moscow, 1826. Reprint, Moscow, 1998.

Ushakova, N. N., and N. A. Figurovskii, *Vasilii Mikhailovich Severgin, 1765-1826 gg*. Moscow, 1981.

Uspenskii, B. A. *Vokrug Trediakovskogo: trudy po istorii russkogo iazyka i russkoi kultur'y*. Moscow, 2008.

Vasetskii, G. S. *Mirovozzrenie M. V. Lomonosova*. Moscow, 1961.

Vasetskii, G. S., and S. R. Mikulinskii, eds. *Izbrannye proizvedeniia russkikh estestvoispytatelei pervoi poloviny XIX veka*. Moscow, 1959.

Vasil'ev, V.N., ed. *Leonard Eiler: K 300-letiiu so dnia rozhdeniia. Sbornik statei*. St. Petersburg, 2008.

Vavilov, S. I. *Mikhail Vasil'evich Lomonosov*. Moscow-Leningrad, 1961.

Velikanova, Olga. *The Public Perception of the Cult of Lenin Based on Archival Materials* (text in Russian). Lewiston, NY: The Edwin Mellen Press, 2001.

Vel'tman, A. F., ed. "Portfel' sluzhebnoi deiatel'nosti Lomonosova." In *Ocherki Rossii*, book 2. Moscow, 1840, 5-85.

Verevkin, M. I. "Zhizn' pokoinogo Mikhaila Vasil'evicha Lomonosova." In *Polnoe sobranie sochinenii Lomonosova*, part 1. St. Petersburg, 1784, III-XVIII.

Voltaire, François Marie Arouet de. *Candide*. Translated by Donald M. Frame. Reprint, New York: New American Library, 1981.

------. *Correspondence and Related Documents, VIII, May 1741-October 1743*. In *Les Ouevres Complètes De Voltaire*, vol. 15, edited by Theodore Besterman et al. Geneva, Institut et Musée Voltaire, 1970.

------. *Eléments de la philosophie de Newton*. In *Les Ouevres Complètes De Voltaire*, vol. 92, edited by W. H. Barber and Ulla Kölvig. Oxford: Voltaire Foundation, 1992.

Vroon, Gail Diane Lenhoff. "The Making of the Medieval Russian Journey." Ph.D. dissertation, University of Michigan, 1978.

Vucinich, Alexander. *Empire of Knowledge: The Academy of Sciences of the USSR (1917-1970)*. Berkeley: University of California Press, 1984.

------. *Science in Russian Culture: A History to 1860*. Stanford: Stanford University Press, 1963.

------. *Science in Russian Culture: 1861-1917*. Stanford: Stanford University Press, 1970.

------. "Soviet Marxism and the History of Science," *Russian Review* 41, no. 2 (April 1982): 123-43.

Wachtel, Andrew. *An Obsession with History: Russian Writers Confront their Past*. Stanford: Stanford University Press, 1994.

Werrett. Simon. "An Odd Sort of Exhibition: The St. Petersburg Academy of Sciences in Enlightened Europe." Ph.D. dissertation, Cambridge University, 2000.

------. *Fireworks: Pyrotechnic Arts and Sciences in European History*. Chicago: University of Chicago Press, 2010.

------. "The Schumacher Affair: Reconfiguring Academic Expertise across Dynasties in Eighteenth-Century Russia." *Osiris* 25, no. 1 (2010): 104-26.

Whittaker, Cynthia H. *The Origins of Modern Russian Education: An Intellectual Biography of Count Sergei Uvarov, 1786-1855*. DeKalb, IL: Northern Illinois University Press, 1984.

Wilberger, Carolyn H. "Voltaire's Russia: Window on the East." In *Studies on Voltaire and the Eighteenth Century*, edited by Theodore Besterman et al., 164. Oxford: Voltaire Foundation, 1976.

Winter, Eduard, ed. *Die deutsch-russische Begegnung und Leonhard Euler: Beiträge zu den beziehungen zwischen der deutschen und der russischen Wissenschaft und Kultur im 18. Jahrhundert*. Berlin: Akademie-Verlag, 1958.

------. "L. Blumentrost d.J. und die Anfänge der Peterburger Akademie der Wissenschaften. Nach Aufzeichnungen von K. F. Svenske." In *Jahrbuch für Geschichte der UdSSR und der volkdemokratischer Länder Europas*, vol. 8, 247-69. Berlin, 1964.

------, ed. *Lomonosov, Schlözer, Pallas: Deutsch-russische Wirtschaftsbeziehungen im 18. Jahrhundert*. Berlin, Akademie-Verlag, 1962.

Winter, Eduard, and A. P. Iushkevich. "O perepiske Leonarda Eilera i G. F. Millera." In *Leonard Eiler: sbornik statei v chest' 250-letiia so dnia rozhdeniia predstavlennykh Akademii nauk SSSR*, edited by M. A. Lavrent'ev, A. P. Iushkevich, and A. T. Grigor'ian, 465-97. Moscow, 1958.

Wirtschafter, Elise Kimerling. *Social Identity in Imperial Russia*. DeKalb, IL: Northern Illinois University Press, 1997.

Wolff, Christian. *Briefe von Christian Wolff aus den Jahren 1719-1753. Ein Beitrag zur Geschichte der Kaiserlichen Academie der Wissenschaften zu St. Petersburg*. St. Petersburg, 1860.

Wortman, Richard S. *Scenarios of Power: Myth and Ceremony in Russian Monarchy*, vol. 1. Princeton: Princeton University Press, 1995.

Yeo, Richard. "Genius, Method and Morality: Images of Newton in Britain, 1760-1860." *Science in Context* 2, no. 2 (Autumn 1988): 257-84.

Zacher, Christian K. *Curiosity and Pilgrimage: The Literature of Discovery in Fourteenth-Century England*. Baltimore: Johns Hopkins University Press, 1976.

Zamkova, V. V. "Fizicheskaia terminologiia v 'Volf'ianskoi eksperi-mental'noi fizike' M. V. Lomonosova." In *Materialy i issledovaniia po leksike Russkogo iazyka XVIII veka*, edited by Iu. S. Sorokin. Moscow-Leningrad, 1965.

Zelov, D. D. *Ofitsial'nye svetskie prazdniki kak iavlenie russkoi kul'tury kontsa XVII—pervoi poloviny XVIII veka*, 2nd ed. Moscow, 2010.

Zhivov, V. M. *Iazyk i kul'tura v Rossii XVIII veka*. Moscow, 1996.

------. "Pervye russkie literaturnye biografii kak sotsial'noe iavlenie: Trediakovskii, Lomonosov, Sumarokov." *Novoe literaturnoe obozrenie* 25 (1997): 24-83.

------. *Razyskaniia v oblasti istorii i predystorii russkoi kul'tury*. Moscow, 2002.

Zhuchkov, V. A., ed. *Khristian Vol'f i filosofiia v Rossii*. St. Petersburg, 2001.

Zubov, V. P. *Istoriografiia estestvennykh nauk v Rossii (XVIII v.–pervaia polovina XIX v.)*. Moscow, 1956.

------. "Lomonosov i slaviano-greko-latinskaia akademiia." *Trudy Instituta istorii estestvoznaniia i tekhniki* 1 (1954): 5-52.

Index

www.ingramcontent.com/pod-product-compliance
Lightning Source LLC
Chambersburg PA
CBHW051102030726
47504CB00006B/1746